Hysteresis of Composites

Edited by Li Longbiao

Published in London, United Kingdom

IntechOpen

Supporting open minds since 2005

Hysteresis of Composites
http://dx.doi.org/10.5772/intechopen.75290
Edited by Li Longbiao

Contributors
Samuel Ratna Kumar P S, John Alexis S, Rajendra Sukhadeorao Dongre, Manvandra Singh, Pushkar Jha, Mulkraj Anand, Rakesh Kumar Gautam, Hüseyin Arslan, Sudipta Sen, Ranjan Kumar Behera, Subash Chandra Mishra, Vladimir Ivchenko, Li Longbiao, Longbiao Li

First published in London, United Kingdom, 2019 by IntechOpen
IntechOpen is the global imprint of INTECHOPEN LIMITED, registered in England and Wales, registration number: 11086078, 7th floor, 10 Lower Thames Street, London, EC3R 6AF, United Kingdom
Printed in Croatia

British Library Cataloguing-in-Publication Data
A catalogue record for this book is available from the British Library

Additional hard and PDF copies can be obtained from orders@intechopen.com

Hysteresis of Composites
Edited by Li Longbiao
p. cm.
Print ISBN 978-1-78984-619-5
Online ISBN 978-1-78984-810-6
eBook (PDF) ISBN 978-1-78984-811-3

An electronic version of this book is freely available, thanks to the support of libraries working with Knowledge Unlatched. KU is a collaborative initiative designed to make high quality books Open Access for the public good. More information about the initiative and links to the Open Access version can be found at www.knowledgeunlatched.org

For EU product safety concerns:
IN TECH d.o.o., Prolaz Marije Krucifikse Kozulić 3, 51000 Rijeka, Croatia, info@intechopen.com or visit our website at intechopen.com.

Meet the editor

Dr. Li Longbiao is a lecturer at the College of Civil Aviation at the Nanjing University of Aeronautics and Astronautics. Dr. Li's research focuses on the fatigue, damage, fracture, reliability, and durability of aircraft and aero engines. In this research area, he is the first author of 130 SCI journal publications, two monographs, three book chapters, 10 Chinese patents, one US patent, and more than 20 refereed conference proceedings. He has been involved in different projects related to structural damage, reliability, and airworthiness design for aircraft and aero engines, supported by the Natural Science Foundation of China, COMAC Company, and the AECC Commercial Aircraft Engine Company.

Contents

Preface

This book introduces the hysteresis and damping of, and damage to, different composites. The hysteresis behavior of fiber-reinforced ceramic–matrix composites (CMCs) is investigated. Based on the interface slip state inside CMCs, the hysteresis loops can be divided into four different cases. The relationship between the internal damage and the hysteresis loops of CMCs is established. Using the experimental hysteresis loops, the fiber/matrix interface frictional coefficient can be obtained. The mechanical hysteresis loops and the fiber/matrix interface frictional coefficient of SiC/CAS and C/SiC composites are predicted using the present models. When the fiber/matrix interface frictional coefficient decreases under cyclic fatigue loading, the fatigue hysteresis loops, fatigue hysteresis dissipated energy, fiber/matrix interface debonding, and slip all change. The fatigue hysteresis dissipated energy first increases and then decreases by decreasing the interface frictional coefficient. The matrix cracking of CMCs is investigated using the energy balance approach. The book also analyzes the relationship between matrix cracking stress, fiber, interface oxidation, and fiber failure, the effects of fiber volume, interface shear stress and interface debonding energy, fiber failure, and oxidation temperature on time-dependent matrix cracking stress, and the experimental matrix cracking stress of different fiber-reinforced CMCs using the present models.

A damping study on aluminum–multiwalled carbon nanotube (MWCNT)-based nanocomposite materials is discussed to increase the damping property for applications like engine heads, pistons, cylinder blocks, and other aerospace components. The Al–MWCNT composites are developed using ball-milling, cold-isostatic, extrusion, and compo-casting processes. Correspondingly, surface morphology and microstructure studies are discussed, as is the improved damping property of Al–MWCNT materials. The hysteresis damping of the composite material is understood by tan delta and storage modulus (or) dynamic Young's modulus. Tan delta and storage modulus of Al–MWCNT composites are performed using a dynamic mechanical analyzer, by varying the temperature range from room temperature to 500°C (heating and cooling phases) at different frequencies. The Al–MWCNT shows good damping properties with increased temperature and a decrease in storage modulus. The mechanisms behind the damping property are also discussed.

The effect of ceramic/graphite addition to the dry sliding wear behavior of copper-based hybrid composites has been assessed at three different normal loads of 9.81, 19.62, and 29.34 N. A wear test is performed by using a pin-on-disc test rig at sliding speeds of 1.30 and 1.84 m/s under ambient conditions. The copper-based hybrid composites are successfully synthesized by using the stir casting route. The samples are characterized by various techniques such as X-ray diffraction, high-resolution-scanning electron microscopy, and a scanning electron microscope equipped with energy-dispersive analysis of X-ray spectroscopy. Microstructural investigations reveal the presence of the WC, TiC, and graphite particles in the hybrid composites. It is observed that a graphite-reinforced hybrid composite shows better wear resistance than a hybrid composite without graphite addition. In all the cases, a cast copper specimen shows the highest wear rate. The observed friction and wear behavior have been explained on the basis of the presence tribofilm on the worn surface. Microstructural characterization of the worn surface and wear debris show that the mechanism of wear is primarily adhesive and oxidative in the case of cast copper, whereas it is a mix of adhesive and abrasive wear in the case of hybrid composites. Wear debris analysis also helps to understand the wear mechanism involved during dry sliding.

I hope this book will help material scientists and engineering designers to understand and master the hysteresis of composites.

Li Longbiao
College of Civil Aviation,
Nanjing University of Aeronautics and Astronautics,
China

Section 1

Hysteresis of Ceramic Matrix Composites

Chapter 1

Hysteresis of Ceramic-Matrix Composites

Li Longbiao

Abstract

In this chapter, the hysteresis behavior of fiber-reinforced ceramic-matrix composites (CMCs) is investigated. Based on the interface slip state inside of CMCs, the hysteresis loops can be divided into four different cases. The relationship between the internal damage and the hysteresis loops of CMCs is established. Using the experimental hysteresis loops, the fiber/matrix interface frictional coefficient can be obtained. The mechanical hysteresis loops and the fiber/matrix interface frictional coefficient of SiC/CAS and C/SiC composites are predicted using the present models. When the fiber/matrix interface frictional coefficient decreases under cyclic fatigue loading, the fatigue hysteresis loops, fatigue hysteresis dissipated energy, fiber/matrix interface debonding, and slip all change. The fatigue hysteresis dissipated energy first increases and then decreases with decreasing interface frictional coefficient.

Keywords: ceramic-matrix composites (CMCs), hysteresis loops, matrix cracking, interface debonding

1. Introduction

Ceramic-matrix composites (CMCs) possess high specific strength and high specific modulus, corrosion, and wear resistance, especially at elevated temperature, and have already been applied on hot section components of commercial aero engine [1, 2].

Upon unloading and subsequent reloading, the fatigue hysteresis loops develop due to the frictional slip that occurred along any interface debonding region [3–6]. Kotil et al. [7] investigated the fatigue hysteresis loops of fiber-reinforced unidirectional CMCs with low and high fiber/matrix interface shear stress. The fatigue hysteresis loops' width decreases with increasing interface shear stress. Pryce and Smith [8] and Keith and Kedward [9] divided the fiber/matrix interface debonding into two cases of partial and complete debonding. Ahn and Curtin [10] investigated the effect of matrix stochastic cracking on the fatigue hysteresis loops of fiber-reinforced unidirectional CMCs. The matrix crack spacing was divided into three cases of long, medium, and short. Vagaggini et al. [11] investigated the fatigue hysteresis loops of fiber-reinforced unidirectional CMCs with weak and strong fiber/matrix interface bonding based on the Hutchinson-Jensen fiber pullout model [12]. Solti et al. [13] investigated the fatigue hysteresis loops of fiber-reinforced unidirectional CMCs when the interface was chemically bonded and partially debonded by adopting the maximum fiber/matrix interface shear strength criterion to determine the fiber/matrix interface debonding length. Li et al. investigated the effects of the fiber/matrix interface debonding [14–16], fiber Poisson contraction [17], fiber fracture [18], and multiple matrix cracking [19] on the fatigue hysteresis loops of fiber-reinforced unidirectional and cross-ply CMCs when the interface was chemically bonded. The area associated with the fatigue hysteresis loops is the energy lost during corresponding unloading/reloading cycles [6]. The fiber/matrix interface shear stress within the debonding region affects the fatigue hysteresis energy lost, i.e., the size and shape of the fatigue hysteresis loops. Cho et al. [20] developed an approach to estimate the interface shear stress from the frictional heating measurements. Li et al. [21] developed an approach to estimate the fiber/matrix interface shear stress of fiber-reinforced unidirectional CMCs from fatigue hysteresis loss energy. However, due to the Poisson contraction effect, the fiber/matrix interface shear stress changes with unloading and reloading [22].

In this chapter, the fatigue hysteresis behavior of fiber-reinforced CMCs is investigated. The relationship between the internal damage and the fatigue hysteresis loops of fiber-reinforced CMCs is established. Using the experimental fatigue hysteresis loops, the fiber/matrix interface frictional coefficient can be obtained. The mechanical fatigue hysteresis loops and the fiber/matrix interface frictional coefficient of SiC/CAS and C/SiC composites are predicted.

2. Theoretical model

Upon loading to the fatigue peak stress, multiple damage mechanisms of the matrix cracking and fiber/matrix interface debonding occur. **Figure 1** shows the unit cell extracted from fiber-reinforced CMCs. The length of the unit cell is half of the matrix crack spacing of $l_c/2$, and the fiber/matrix interface debonding length is l_d. The fiber and the matrix radius are r_f and R.

Based on the fiber/matrix interface slip mechanism, the fatigue hysteresis loops of fiber-reinforced CMCs can be divided into four different cases, as the following:

1. Case I, the fiber complete sliding relative to the matrix in the partial debonding interface between the fiber and the matrix.

2. Case II, the fiber partial sliding relative to the matrix in the partial debonding interface between the fiber and the matrix.

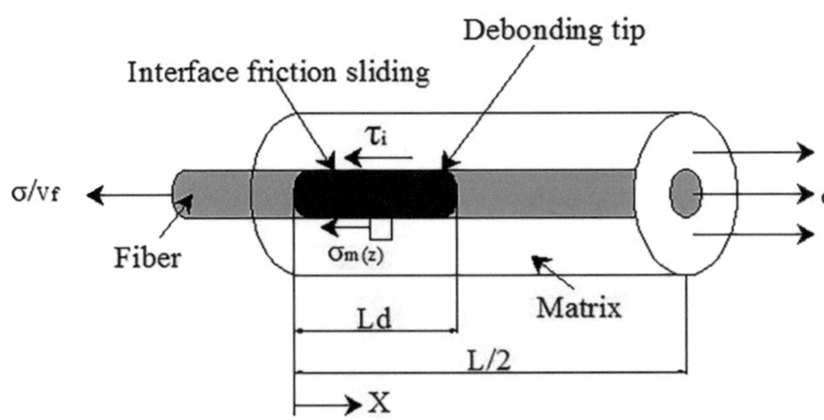

Figure 1.
The unit cell of composite.

3. Case III, the fiber partial sliding relative to the matrix in the complete debonding interface between the fiber and the matrix.

4. Case IV, the fiber complete sliding relative to the matrix in the complete debonding interface between the fiber and the matrix.

2.1 Case I

Upon unloading to the applied stress of σ (i.e., $\sigma_{\min} < \sigma < \sigma_{\max}$), the fiber/matrix interface debonding region can be divided into two regions of the interface counter slip region and interface slip region. When the applied stress is higher than the unloading transition stress of $\sigma > \sigma_{tr\_pu}$, the fiber axial stress distribution upon unloading can be determined using the following equation:

$$\sigma_f(z) = \begin{cases} \dfrac{\sigma}{V_f} - \dfrac{\alpha v_f}{V_f(\alpha v_f + \gamma v_m)}(\overline{\sigma} - \sigma)\left(e^{-\lambda z} - 1\right), z \in [0, y] \\ \dfrac{\sigma}{V_f} - \dfrac{\alpha v_f}{V_f(\alpha v_f + \gamma v_m)}(\overline{\sigma} - \sigma)\left(e^{\lambda(z-2y)} - 1\right), z \in [y, l_d] \\ \left(\dfrac{\sigma}{V_f} - \dfrac{\alpha v_f(\overline{\sigma} - \sigma)}{V_f(\alpha v_f + \gamma v_m)}\left(e^{\lambda(L_d-2y)} - 1\right) - \sigma_{fo}\right)e^{-\frac{\rho(z-L_d)}{r_f}} + \sigma_{fo}, z \in \left[l_d, \dfrac{l_c}{2}\right] \end{cases} \tag{1}$$

where

$$y = \frac{1}{2}\left\{l_d(\sigma_{\max}) - \frac{1}{\lambda}\ln\left[1 + \frac{(\alpha v_f + \gamma v_m)(\sigma - \sigma_0)}{\alpha v_f(1+\beta)(\overline{\sigma} - \sigma)}\right]\right\} \tag{2}$$

When the applied stress is lower than the unloading transition stress of $\sigma_{\min} < \sigma < \sigma_{tr\_pu}$, the fiber/matrix interface counter slip length occupies the entire interface debonding region, and the stress within the fiber can be determined using Eq. (1) by setting $y = l_d$.

Upon reloading to the applied stress of σ (i.e., $\sigma_{\min} < \sigma < \sigma_{\max}$), the fiber/matrix interface debonding region can be divided into three regions of new slip region, counter slip region, and slip region. When the applied stress is higher than the reloading transition stress of $\sigma < \sigma_{tr\_pr}$, the fiber axial stress distribution upon reloading can be determined using the following equation:

$$\sigma_f(z) = \begin{cases} \dfrac{\sigma}{V_f} - \dfrac{\alpha v_f}{V_f(\alpha v_f + \gamma v_m)}(\overline{\sigma} - \sigma)\left(e^{\lambda z} - 1\right), z \in [0, z'] \\ \dfrac{\sigma}{V_f} - \dfrac{\alpha v_f}{V_f(\alpha v_f + \gamma v_m)}(\overline{\sigma} - \sigma)\left(e^{-\lambda(z-2z')} - 1\right), z \in [z', y] \\ \dfrac{\sigma}{V_f} - \dfrac{\alpha v_f}{V_f(\alpha v_f + \gamma v_m)}(\overline{\sigma} - \sigma)\left(e^{\lambda(z-2y+2z')} - 1\right), z \in [y, l_d] \\ \left(\dfrac{\sigma}{V_f} - \dfrac{\alpha v_f(\overline{\sigma} - \sigma)}{V_f(\alpha v_f + \gamma v_m)}\left(e^{\lambda(l_d-2y+2z')} - 1\right) - \sigma_{fo}\right)e^{-\frac{\rho(z-l_d)}{r_f}} + \sigma_{fo}, z \in [l_d, l_c/2] \end{cases} \tag{3}$$

where

$$z' = y(\sigma_{\min}) - \frac{1}{2}\left\{l_d(\sigma_{\max}) - \frac{1}{\lambda}\ln\left[1 + \frac{(\alpha v_f + \gamma v_m)(\sigma - \sigma_0)}{\alpha v_f(\overline{\sigma} - \sigma)(1+\beta)}\right]\right\} \tag{4}$$

When the applied stress is higher than the transition stress of $\sigma_{tr\_pr} < \sigma < \sigma_{max}$, the new slip length occupies the entire interface debonding length, and the stress within the fiber can be determined using Eq. (3) by setting $z' = l_d$.

2.2 Case II

For the interface slip Case II, the fiber complete slides relative to the matrix in the partial debonding interface between the fiber and the matrix. The interface counter slip length y upon complete unloading is less than the interface debonding length of l_d, i.e., $y(\sigma_{min}) < l_d$. The fiber axial stress distribution can be determined by Eq. (1), and the interface counter slip length y is determined by Eq. (2). The new interface slip length of z' upon reloading to the peak stress of σ_{max} is less than the interface debonding length of l_d, i.e., $z'(\sigma_{max}) < l_d$. The fiber axial stress distribution is determined by Eq. (3); the new interface slip length z' is determined by Eq. (4).

2.3 Case III

For the interface slip Case III, the fiber partial slides relative to the matrix in the complete debonding interface between the fiber and the matrix. Upon complete unloading, the interface counter slip length of y is less than the half matrix crack space of $l_c/2$, i.e., $y(\sigma_{min}) < l_c/2$. The fiber axial stress distribution upon unloading can be determined using the following equation:

$$\sigma_f(z) = \begin{cases} \dfrac{\sigma}{V_f} - \dfrac{\alpha v_f}{V_f(\alpha v_f + \gamma v_m)}(\overline{\sigma} - \sigma)\left(e^{-\lambda z} - 1\right), z \in [0, y] \\ \dfrac{\sigma}{V_f} - \dfrac{\alpha v_f}{V_f(\alpha v_f + \gamma v_m)}(\overline{\sigma} - \sigma)\left(e^{\lambda(z-2y)} - 1\right), z \in [y, l_c/2] \end{cases} \tag{5}$$

where

$$y = \frac{1}{2\lambda}\ln\frac{\alpha v_f(1+\beta)(\overline{\sigma} - \sigma)(\overline{\sigma} - \sigma_{max}) + (\alpha v_f + \gamma v_m)(\overline{\sigma} - \sigma)\sigma_{max}}{\alpha v_f(1+\beta)(\overline{\sigma} - \sigma)(\overline{\sigma} - \sigma_{max}) + (\alpha v_f + \gamma v_m)(\overline{\sigma} - \sigma_{max})\sigma} \tag{6}$$

Upon reloading to the peak stress of σ_{max}, the interface new slip length of z' is less than the half matrix crack space of $l_c/2$, i.e., $z'(\sigma_{min}) < l_c/2$. The fiber axial stress distribution upon reloading can be determined using the following equation:

$$\sigma_f(z) = \begin{cases} \dfrac{\sigma}{V_f} - \dfrac{\alpha v_f}{V_f(\alpha v_f + \gamma v_m)}(\overline{\sigma} - \sigma)\left(e^{\lambda z} - 1\right), z \in [0, z'] \\ \dfrac{\sigma}{V_f} - \dfrac{\alpha v_f}{V_f(\alpha v_f + \gamma v_m)}(\overline{\sigma} - \sigma)\left(e^{-\lambda(z-2z')} - 1\right), z \in [z', y] \\ \dfrac{\sigma}{V_f} - \dfrac{\alpha v_f}{V_f(\alpha v_f + \gamma v_m)}(\overline{\sigma} - \sigma)\left(e^{\lambda(z-2y+2z')} - 1\right), z \in [y, l_c/2] \end{cases} \tag{7}$$

where

$$z' = \frac{1}{2\lambda}\ln\frac{\alpha v_f(1+\beta)(\overline{\sigma} - \sigma)(\overline{\sigma} - \sigma_{max}) + (\alpha v_f + \gamma v_m)(\overline{\sigma} - \sigma_{max})\sigma}{\alpha v_f(1+\beta)(\overline{\sigma} - \sigma_{min})(\overline{\sigma} - \sigma_{max}) + (\alpha v_f + \gamma v_m)(\overline{\sigma} - \sigma_{max})\sigma_{min}} \tag{8}$$

2.4 Case IV

For the interface slip Case IV, the fiber complete slides relative to the matrix in the complete debonding interface between the fiber and the matrix. Upon unloading to the transition stress of σ_{tr_fu}, the interface counter slip length of y reaches the half matrix crack space of $l_c/2$. When the applied stress of $\sigma > \sigma_{tr\_fu}$, the interface counter slip length y is less than the half matrix crack space of $l_c/2$. The fiber axial stress distribution is determined by Eq. (5), and the interface counter slip length y is determined by Eq. (6). When the applied stress is lower than the transition stress of $\sigma_{min} < \sigma < \sigma_{tr\_fu}$, the interface counter slip occupies the entire matrix crack space, and the fiber axial stress distribution is determined by Eq. (5) by setting $y = l_c/2$.

Upon reloading to the transition stress of σ_{tr_fr}, the new interface slip length z' reaches the half matrix crack space of $l_c/2$. When $\sigma < \sigma_{tr\_fr}$, the interface new slip length z' is less than the half matrix crack space $l_c/2$. The fiber axial stress distribution is determined by Eq. (7), and the interface new slip length z' is determined by Eq. (8). When the applied stress is higher than the transition stress of $\sigma_{tr\_fr} < \sigma < \sigma_{max}$, new slip length occurs over the entire matrix crack space, and the fiber axial stress distribution is given by Eq. (7) by setting $z' = l_c/2$.

2.5 Hysteresis loops and hysteresis dissipated energy

For the interface slip Case II, the unloading and reloading stress–strain relationships are given by the following equation:

$$\begin{aligned}\varepsilon_{c\_pu} = &\frac{\sigma}{V_f E_f} - \frac{2\alpha v_f(\overline{\sigma} - \sigma)}{\lambda V_f E_f l_c(\alpha v_f + \gamma v_m)}\left(1 - 2e^{-\lambda y} + e^{\lambda(l_d - 2y)}\right) \\ &+ \frac{\alpha v_f(\overline{\sigma} - \sigma)}{V_f E_f(\alpha v_f + \gamma v_m)}\left[1 - e^{\lambda(l_d - 2y)}(1 - 2l_d/l_c)\right]\end{aligned} \quad (9)$$

$$\begin{aligned}\varepsilon_{c\_pr} = &\frac{\sigma}{V_f E_f} - \frac{4\alpha v_f(\overline{\sigma} - \sigma)}{\lambda V_f E_f l_c(\alpha v_f + \gamma v_m)}\left(e^{\lambda z\prime} - e^{-\lambda(y - 2z\prime)} + e^{\lambda(l_d - 2y + 2z\prime)} - 1\right) \\ &+ \frac{\alpha v_f(\overline{\sigma} - \sigma)}{V_f E_f(\alpha v_f + \gamma v_m)}\left[1 - e^{\lambda(l_d - 2y + 2z\prime)}(1 - 2l_d/l_c)\right]\end{aligned} \quad (10)$$

For the interface slip Case I, the unloading stress–strain relationship can be divided into two regions. When the applied stress is higher than the transition stress of $\sigma > \sigma_{tr\_pu}$, the unloading strain is determined by Eq. (9), and when the applied stress is lower than the transition stress of $\sigma < \sigma_{tr\_pu}$, the unloading strain is determined by Eq. (9) by setting $y = l_d$. The reloading stress–strain relationship is divided into two regions. When the applied stress is lower than the transition stress of $\sigma < \sigma_{tr\_pr}$, the reloading strain is determined by Eq. (10), and when the applied stress is higher than the transition stress of $\sigma > \sigma_{tr\_pr}$, the reloading strain is determined by Eq. (10) by setting $z' = l_d$.

For the interface slip Case III, the unloading and reloading stress–strain relationships are determined by the following equation:

$$\varepsilon_{c\_fu} = \frac{\sigma}{V_f E_f} + \frac{\alpha v_f(\overline{\sigma} - \sigma)}{V_f E_f(\alpha v_f + \gamma v_m)} - \frac{2\alpha v_f(\overline{\sigma} - \sigma)}{\lambda V_f E_f l_c(\alpha v_f + \gamma v_m)}\left(1 - 2e^{-\lambda y} + e^{\lambda(l_c/2 - 2y)}\right) \quad (11)$$

$$\begin{aligned}\varepsilon_{c\_fr} = &\frac{\sigma}{V_f E_f} + \frac{\alpha v_f(\overline{\sigma} - \sigma)}{V_f E_f(\alpha v_f + \gamma v_m)} \\ &- \frac{2\alpha v_f(\overline{\sigma} - \sigma)}{\lambda V_f E_f l_c(\alpha v_f + \gamma v_m)}\left(2e^{\lambda z\prime} - 2e^{-\lambda(y - 2z\prime)} + e^{\lambda(l_c/2 - 2y + 2z\prime)} - 1\right)\end{aligned} \quad (12)$$

For the interface slip Case IV, the unloading stress–strain relationship is divided into two regions. When the applied stress is higher than the transition stress of $\sigma > \sigma_{tr\_fu}$, the unloading strain is determined by Eq. (11), and when the applied stress is lower than the transition stress of $\sigma < \sigma_{tr\_fu}$, the unloading strain is determined by Eq. (11) by setting $y = l_c/2$. The reloading stress–strain relationship is divided into two regions. When the applied stress is lower than the transition stress of $\sigma < \sigma_{tr\_fr}$, the reloading strain is determined by Eq. (12); when the applied stress is higher than the transition stress of $\sigma > \sigma_{tr\_fr}$, the reloading strain is determined by Eq. (12) by setting $z' = l_c/2$.

Under cyclic fatigue loading, the area associated with the fatigue hysteresis loops is the energy lost during the corresponding cycle, which is defined by the following equation:

$$U = \int_{\sigma_{\min}}^{\sigma_{\max}} [\varepsilon_{cu}(\sigma) - \varepsilon_{cr}(\sigma)] d\sigma \tag{13}$$

3. Result and discussion

The experimental and predicted fatigue hysteresis loops and interface slip of fiber-reinforced unidirectional SiC/CAS composite under the fatigue peak stress of σ_{max} = 185 MPa are shown in **Figure 2**. Upon unloading and reloading, the unloading interface counter slip length approaches the interface debonding length at the unloading transition stress, and the reloading interface new slip length approaches the interface debonding length at the reloading transition stress. The fatigue unloading/reloading hysteresis loops under σ_{max} = 185 MPa correspond to the interface slip Case I.

The experimental and predicted fatigue hysteresis loops and interface slip of fiber-reinforced unidirectional SiC/CAS composite under the fatigue peak stress of σ_{max} = 242 MPa are shown in **Figure 3**. Upon unloading the and reloading, the unloading interface counter slip length and the reloading interface new slip length do not approach the interface debonding length. The fatigue hysteresis loops under the fatigue peak stress of σ_{max} = 242 MPa correspond to the interface slip Case II.

The experimental and predicted fatigue hysteresis loop and interface slip of fiber-reinforced unidirectional SiC/CAS composite under the fatigue peak stress of σ_{max} = 300 MPa are shown in **Figure 4**. The interface partial debonds and the fiber partial slides upon unloading and reloading. The unloading interface counter slip length and reloading interface new slip length do not approach the interface debonding length. The fatigue hysteresis loops under the fatigue peak stress of σ_{max} = 300 MPa correspond to the interface slip Case III.

The experimental and predicted fatigue hysteresis loops and interface slip of SiC/CAS composite under the fatigue peak stress of σ_{max} = 350 MPa are shown in **Figure 5**. The interface complete debonds and the fiber complete slides upon unloading and reloading. The unloading interface counter slip length and the new interface slip length upon reloading approach the interface debonding length. The fatigue hysteresis loops under the fatigue peak stress of σ_{max} = 350 MPa correspond to the interface slip Case IV.

The fatigue hysteresis loops as a function of the fiber/matrix interface frictional coefficient of fiber-reinforced unidirectional SiC/CAS composite under the fatigue peak stress of σ_{max} = 185 MPa are shown in **Figure 6(a)**. The fatigue hysteresis dissipated energy as a function of the fiber/matrix interface frictional coefficient of fiber-reinforced unidirectional SiC/CAS composite is shown in **Figure 6(b)**. When the interface frictional coefficient is μ = 0.05~0.3, the fatigue hysteresis dissipated energy increases with the decreasing fiber/matrix interface frictional coefficient, and the interface partial debonds ($l_d < l_c/2$ in **Figure 6(c)**), and the fiber complete

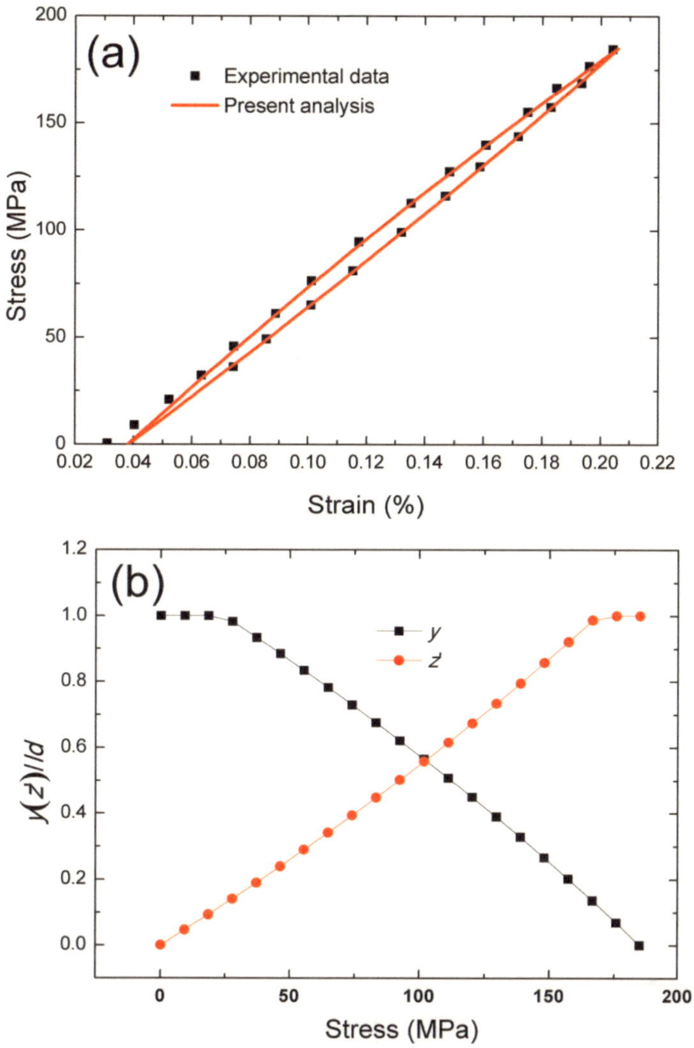

Figure 2.
(a) The experimental and predicted fatigue hysteresis loops and (b) the fiber/matrix interface counter slip length and the interface new slip length versus the applied stress of fiber-reinforced unidirectional SiC/CAS composite under the fatigue peak stress of σ_{max} *= 185 MPa.*

slides relative to the matrix in the interface debonding region (i.e., $y(\sigma_{min}) = l_d$ in **Figure 6(d)**). The fatigue hysteresis loops correspond to the interface slip Case I. When the fiber/matrix interface frictional coefficient is $\mu = 0.01{\sim}0.05$, the fatigue hysteresis dissipated energy increases to the peak value and decreases with the decreasing interface frictional coefficient, and the interface complete debonds ($l_d = l_c/2$ in **Figure 6(c)**), and the fiber complete slides relative to the matrix in the interface debonding region ($y(\sigma_{min}) = l_d$ in **Figure 6(d)**). The fatigue hysteresis loops correspond to the interface slip Case IV.

4. Experimental comparison

The fatigue hysteresis loops, fatigue hysteresis dissipated energy, and interface slip of unidirectional C/SiC composite at room and elevated temperatures corresponding to different number of applied cycles are analyzed.

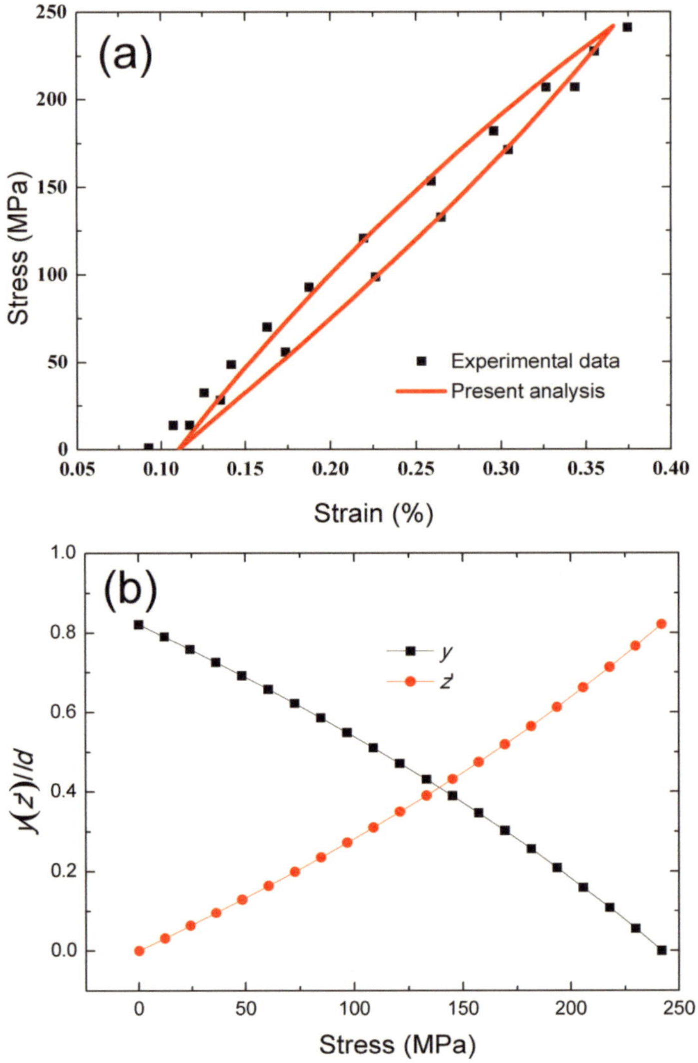

Figure 3.
(a) The experimental and predicted fatigue hysteresis loops and (b) the fiber/matrix interface counter slip length and the interface new slip length versus the applied stress of fiber-reinforced unidirectional SiC/CAS composite under the fatigue peak stress of σ_{max} *= 242 MPa.*

4.1 Room temperature

The fatigue hysteresis dissipated energy as a function of the fiber/matrix interface frictional coefficient of C/SiC composite under the fatigue peak stress of $\sigma_{\max}$ = 140 MPa is shown in **Figure 7**. The fatigue hysteresis dissipated energy increases with decreasing fiber/matrix interface frictional coefficient to the peak value of U = 22 kJ/m^3 and then decreases with the fiber/matrix interface frictional coefficient to U = zero kJ/m^3. The experimental fatigue hysteresis dissipated energy is U = 18, 7.7, 6.5, 6.1, 4.2, 4, 3.8, and 3.4 kJ/m^3 corresponding to N = 1, 15,155, 58,804, 139,326, 234,783, 665,129, 816,908, and 1,005,541, respectively; the corresponding fiber/matrix interface frictional coefficient is $\mu = 1.1 \times 10^{-2}$, 1.5×10^{-3}, 1.3×10^{-3}, 1.1×10^{-3}, 8.4×10^{-4}, 8×10^{-4}, 7.5×10^{-4}, and 6.8×10^{-4}, respectively. Under the fatigue peak stress of $\sigma_{\max}$ = 140 MPa, the fatigue hysteresis

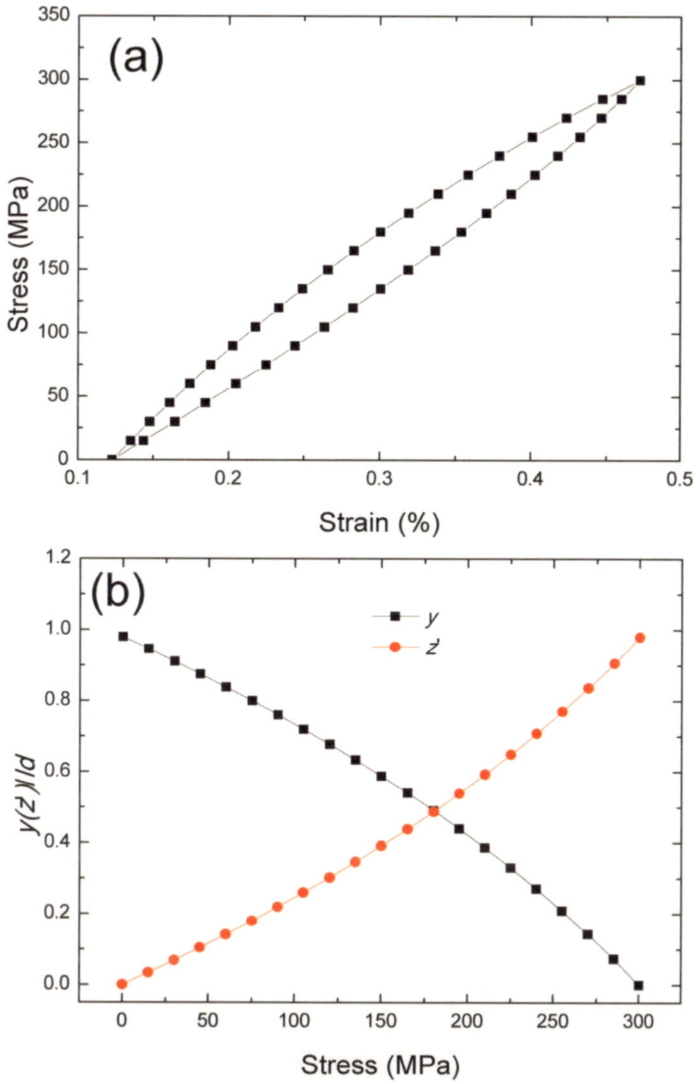

Figure 4.
(a) The experimental and predicted fatigue hysteresis loops and (b) the interface counter slip length and the interface new slip length versus the applied stress of fiber-reinforced unidirectional SiC/CAS composite under the fatigue peak stress of σ_{max} = 300 MPa.

dissipated energy at N = 1 lies in the right part of the fatigue hysteresis dissipated energy versus the fiber/matrix interface frictional coefficient curve. The fatigue hysteresis loops at N = 1 correspond to the interface slip Case II. When the applied cycles increase, the fiber/matrix interface frictional coefficient decreases. The fatigue hysteresis loops at N = 15,155 correspond to the interface slip Case IV.

4.2 Elevated temperature

The fatigue hysteresis dissipated energy as a function of the fiber/matrix interface frictional coefficient under the fatigue peak stress of σ_{max} = 180 MPa is shown in **Figure 8**. The fatigue hysteresis dissipated energy first increases with decreasing interface frictional coefficient to the peak value of U = 36.5 kJ/m^3 and then

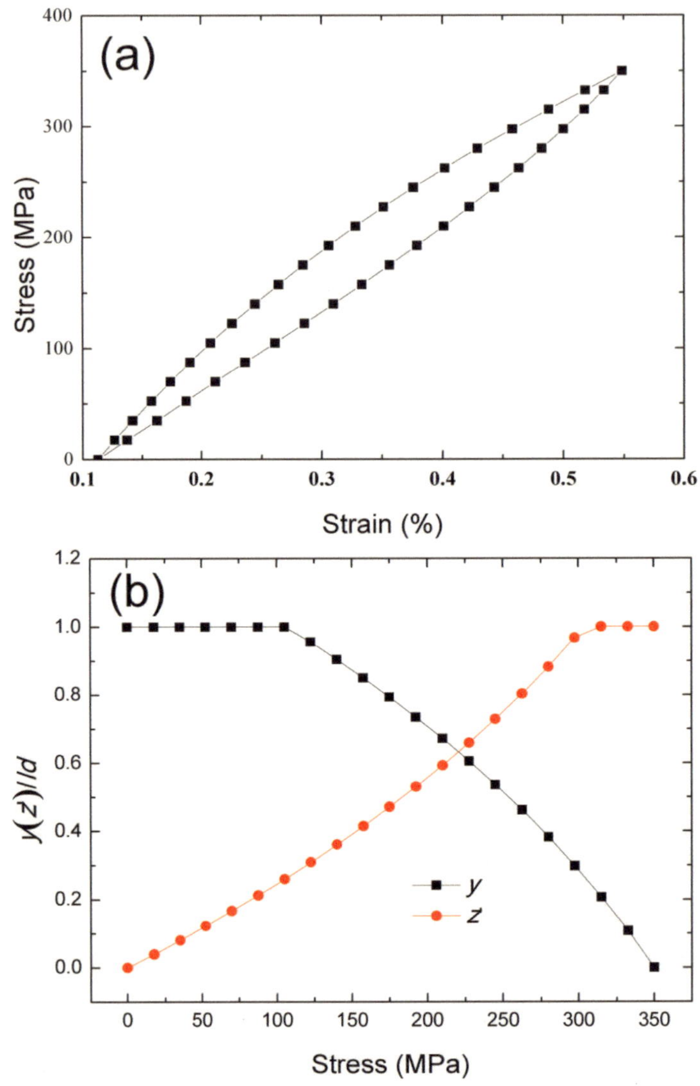

Figure 5.
(a) The experimental and predicted fatigue hysteresis loops and (b) the interface counter slip length and the interface new slip length versus the applied stress of fiber-reinforced unidirectional SiC/CAS composite under the fatigue peak stress of σ_{max} = 350 MPa.

decreases with the interface frictional coefficient to U = zero kJ/m^3. The experimental fatigue hysteresis dissipated energy is U = 30, 22, 15, 9.8, 9.3, 8.7, 8, 7.5, 6.9, 6.7, 6.1, 6, and 5.8 kJ/m^3, corresponding to N = 1, 5, 10, 100, 1000, 8000, 21,000, 36,000, 55,000, 65,000, 75,000, 85,000, and 87,000, respectively; the corresponding fiber/matrix interface frictional coefficient is $\mu = 1.1 \times 10^{-2}$, 2.7×10^{-3}, 1.8×10^{-3}, 1.2×10^{-3}, 1.1×10^{-3}, 1×10^{-3}, 9×10^{-4}, 8.5×10^{-4}, 8×10^{-4}, 7.8×10^{-4}, 5.1×10^{-4}, 5×10^{-4}, and 4.9×10^{-4}, respectively. Under the fatigue peak stress of σ_{max} = 180 MPa, the fatigue hysteresis dissipated energy at N = 1 lies in the right part of the fatigue hysteresis dissipated energy versus the interface frictional coefficient curve. The fatigue hysteresis loops at N = 1 correspond to the interface slip Case II. When the applied cycles increase, the fiber/matrix interface frictional coefficient decreases due to the oxidation of the interphase. The fatigue hysteresis loops at N = 5 correspond to the interface slip Case IV.

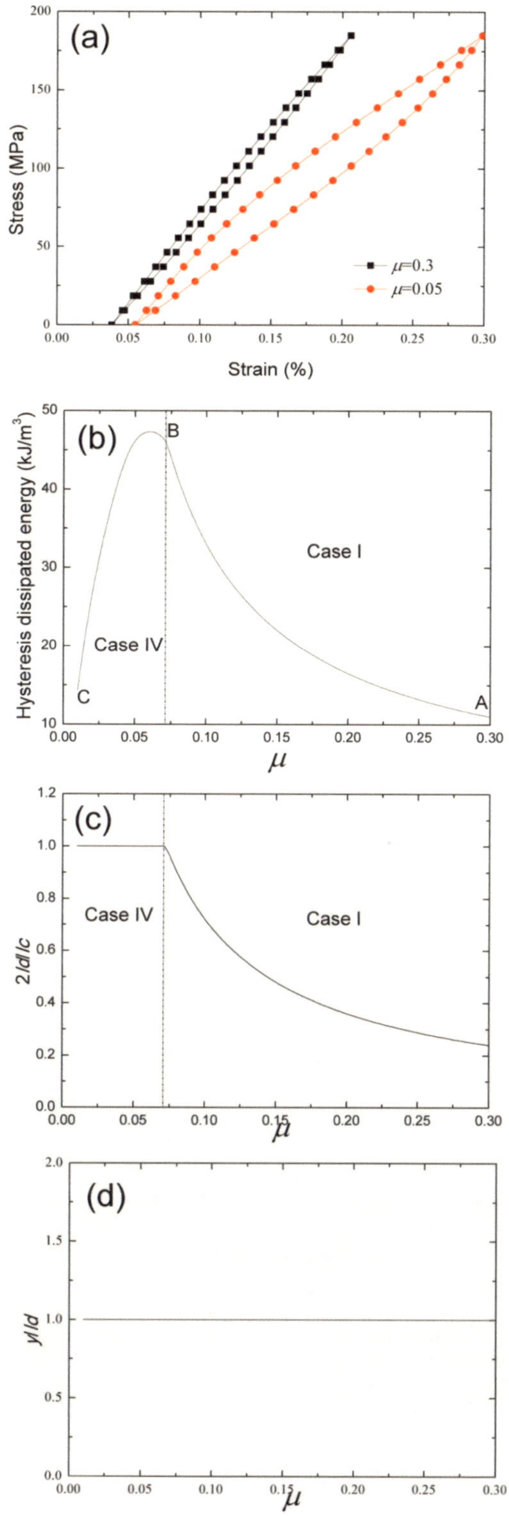

Figure 6.
(a) The fatigue hysteresis loops corresponding to different fiber/matrix interfacial frictional coefficient; (b) the fatigue hysteresis dissipated energy versus the fiber/matrix interfacial frictional coefficient curves; (c) the interface debonding length versus the interfacial frictional coefficient curves; and (d) the interface counter slip length versus the interfacial frictional coefficient curves of fiber-reinforced unidirectional SiC/CAS composite under the fatigue peak stress of σ_{max} = 185 *MPa.*

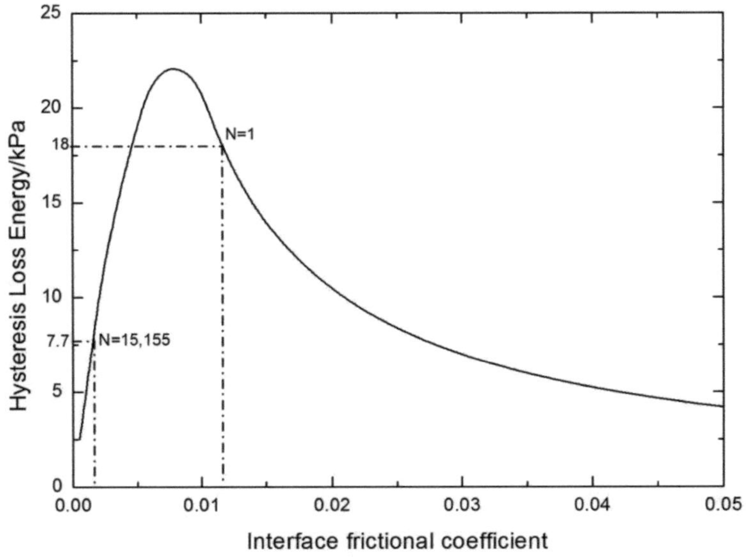

Figure 7.
The experimental and predicted fatigue hysteresis dissipated energy versus the fiber/matrix interface frictional coefficient of fiber-reinforced unidirectional C/SiC composite under the fatigue peak stress of σ_{max} = *140 MPa at room temperature.*

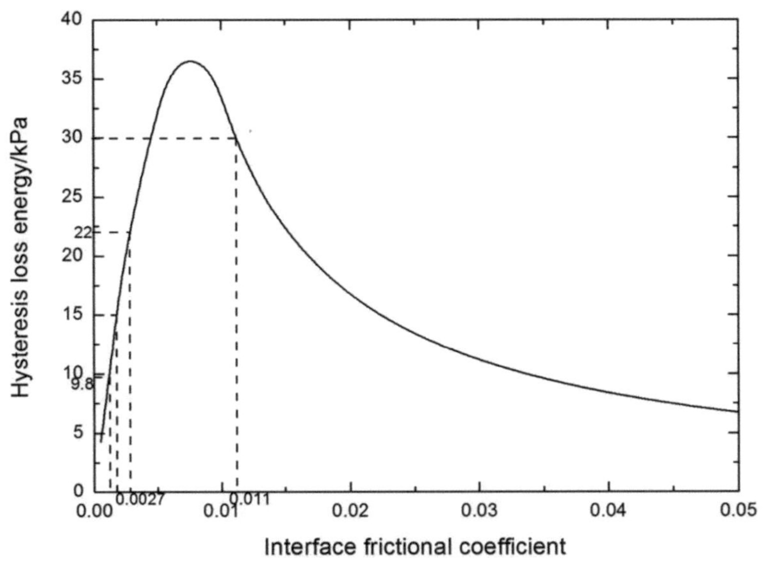

Figure 8.
The experimental and predicted fatigue hysteresis dissipated energy versus the fiber/matrix interface frictional coefficient of fiber-reinforced unidirectional C/SiC composite under the fatigue peak stress of σ_{max} = *180 MPa at 800°C in air atmosphere.*

5. Conclusion

In this chapter, the fatigue hysteresis behavior of fiber-reinforced CMCs was investigated. The relationship between the internal damage and the fatigue hysteresis loops of fiber-reinforced CMCs was established. Using the experimental fatigue hysteresis loops, the fiber/matrix interface frictional coefficient corresponding to different applied cycles can be obtained. The mechanical fatigue hysteresis loops and the fiber/matrix interface frictional coefficient of SiC/CAS and C/SiC composites were predicted. The fatigue hysteresis loops of fiber-reinforced CMCs can be divided into four different cases, as the following:

1. Case I, the fiber complete sliding relative to the matrix in the partial debonding interface between the fiber and the matrix.

2. Case II, the fiber partial sliding relative to the matrix in the partial debonding interface between the fiber and the matrix.

3. Case III, the fiber partial sliding relative to the matrix in the complete debonding interface between the fiber and the matrix.

4. Case IV, the fiber complete sliding relative to the matrix in the complete debonding interface between the fiber and the matrix.

Acknowledgements

The work reported here is supported by the Fundamental Research Funds for the Central Universities (Grant No. NS2019038).

Author details

Li Longbiao
College of Civil Aviation, Nanjing University of Aeronautics and Astronautics, Nanjing, China

*Address all correspondence to: llb451@nuaa.edu.cn

References

[1] Li L. Damage, Fracture and Fatigue of Ceramic-Matrix Composites. Singapore: Springer Nature Singapore Pte Ltd.; 2018. DOI: 10.1007/978-981-13-1783-5. ISBN: 978-981-13-1782-8

[2] Li L. Thermomechanical Fatigue of Ceramic-Matrix Composites. New Jersey, United States: Wiley-VCH; 2019. Available from: https://onlinelibrary.wiley.com/doi/book/10.1002/9783527822614. ISBN: 978-3-527-34637-0

[3] Li L. Synergistic effects of temperature, oxidation, loading frequency and stress-rupture on damage evolution of cross-ply ceramic-matrix composites under cyclic fatigue loading at elevated temperatures in oxidizing atmosphere. Engineering Fracture Mechanics. 2017;**175**:15-30

[4] Li L. Synergistic effects of stress-rupture and cyclic loading on strain response of fiber-reinforced ceramic-matrix composites at elevated temperature in oxidizing atmosphere. Materials. 2017;**10**:182

[5] Li L. Damage evolution of cross-ply ceramic-matrix composites under stress-rupture and cyclic loading at elevated temperatures in oxidizing atmosphere. Materials Science and Engineering A. 2017;**688**:315-321

[6] Fantozzi G, Reynaud P. Mechanical hysteresis in ceramic matrix composites. Materials Science and Engineering A. 2009;**521-522**:18-23

[7] Kotil T, Holmes JW, Comninou M. Origin of hysteresis observed during fatigue of ceramic-matrix composites. Journal of the American Ceramic Society. 1990;**73**:1879-1883

[8] Pryce AW, Smith PA. Matrix cracking in unidirectional ceramic matrix composites under quasi-static and cyclic loading. Acta Metallurgica et Materialia. 1993;**41**:1269-1281

[9] Keith WP, Kedward KT. The stress-strain behavior of a porous unidirectional ceramic matrix composite. Composites. 1995;**26**:163-174

[10] Ahn BK, Curtin WA. Strain and hysteresis by stochastic matrix cracking in ceramic matrix composites. Journal of the Mechanics and Physics of Solids. 1997;**45**:177-209

[11] Vagaggini E, Domergue JM, Evans AG. Relationships between hysteresis measurements and the constituent properties of ceramic matrix composites: I, theory. Journal of the American Ceramic Society. 1995;**78**:2709-2720

[12] Hutchison JW, Jensen HM. Models of fiber debonding and pullout in brittle composites with friction. Mechanics of Materials. 1990;**9**:139-163

[13] Solti JP, Robertson DD, Mall S. Estimation of interfacial properties from hysteresis energy loss in unidirectional ceramic matrix composites. Advanced Composite Materials. 2000;**9**(3):161-173

[14] Li L. Modeling for cyclic loading/unloading hysteresis loops of carbon fiber-reinforced ceramic-matrix composites at room and elevated temperatures. Part I: Theoretical analysis. Engineering Fracture Mechanics. 2016;**164**:117-136

[15] Li L. Modeling for cyclic loading/unloading hysteresis loops of carbon fiber-reinforced ceramic-matrix composites at room and elevated temperatures. Part II: Experimental comparisons. Engineering Fracture Mechanics. 2016;**164**:137-154

[16] Li L. Hysteresis loops of carbon fiber-reinforced ceramic-matrix composites with different fiber preforms. Ceramics International. 2016; **42**:16535-16551

[17] Li L, Song Y. Estimate interface frictional coefficient of ceramic matrix composites from hysteresis loops. Journal of Composite Materials. 2011;**45**: 989-1006

[18] Li L, Song Y, Sun Y. Modeling loading/unloading hysteresis behavior of unidirectional C/SiC ceramic matrix composites. Applied Composite Materials. 2013;**20**:655-672

[19] Li L. Fatigue hysteresis behavior of cross-ply C/SiC ceramic matrix composites at room and elevated temperatures. Materials Science and Engineering A. 2013;**586**:160-170

[20] Cho C, Holmes JW, Barber JR. Estimation of interfacial shear in ceramic composites from frictional heating measurements. Journal of the American Ceramic Society. 1991;**74**: 2802-2808

[21] Li L. Assessment of the interfacial properties from fatigue hysteresis loss energy in ceramic-matrix composites with different fiber preforms at room and elevated temperatures. Materials Science and Engineering A. 2014;**613**: 17-36

[22] Li L. Fatigue hysteresis behavior of unidirectional C/SiC ceramic matrix composites at room and elevated temperatures. Materials Science and Engineering A. 2015;**625**:1-18

Section 2

Damping of Composites

Chapter 2

Damping Study on MWCNT-Reinforced Al Composites

Paul Suresh Samuel Ratna Kumar and Savariar John Alexis

Abstract

In this chapter, damping study on Aluminium-MWCNT based nanocomposites materials are discussed to increase the damping property for applications like engine head, pistons and cylinder blocks and other aerospace components. The Al-MWCNT composites were developed using Ball-milling, cold-isostatic, extrusion and compo-casting process. Corresponding, surface morphology and micro-structure study were discussed. The improved damping property of Al-MWCNT materials were discussed. The hysteresis damping of the composite material was understood by Tan delta and storage modulus (or) dynamic young's modulus. Tan delta and storage modulus of Al-MWCNT composites were performed using dynamic mechanical analyser, by varying the temperature range from room temperature to 500°C (heating phases and cooling phases), at different frequencies. The Al-MWCNT shows some good damping property with increased temperature and decrease in storage modulus. The mechanisms behind the damping property were discussed.

Keywords: aluminium (Al), multi-wall carbon nanotube (MWCNT), damping property, storage modulus, temperature, frequency

1. Introduction

In various structural applications like aerospace, aeronautical, marine and automotive cause vibrations occur during their regular operation after a period of usage which affects the human comfort due to their unwanted noise [1]. To overcome the vibration problems, researchers have proposed newly developed materials with high damping capacity [2, 3]. These vibrations are also prone to increase with temperature.

In recent years, advanced materials like nanocomposites, have been developed to play a vital role in component design with good damping, stiffness and low density. Metal matrix nanocomposites (MMCs) are one of the advanced composite materials which combines nano-based reinforcements in a metal-based matrix. They have excellent properties for application-based design [3].

Many industries use aluminium and its alloys due to their low density/weight and high mechanical properties. However, the damping property of the aluminium and its alloys is poor. Further to develop the damping of the material aluminium based nanocomposites were developed based on the requirements. Many nano-based reinforcements were used to improve the strength and stiffness properties, still they suffer from limitations like poor damping and plasticity. For the above reasons carbon nanotubes (CNT) were used as reinforcements in proper composition to transfer the physical and mechanical properties into the bulk aluminium material [4].

Nanotubes have high thermal and electrical conductivity, strength and modulus, and also it shows its potential in damping applications due to their low density and high specific surface area. Very few studies have been carried out on aluminium reinforced CNT composites for improving the damping property. Many researchers used CNT in polymer matrix composites to improve the damping property [5, 6]. Based on the above studies, CNT with proper composition/dispersion can improve the mechanical strength and damping property. The positive result is due to friction and slippage that occurs in the inner and outer walls of carbon nanotubes and polymer matrix [7, 8]. The mechanism behind the improved damping property is due to interfacial slip on CNT reinforced composites. CNT reinforced composites have good damping property which is influenced by weight (or) volume fraction, dispersion pattern of nanotubes, temperature and frequency [4–6, 8].

Only very few research works have been carried out to find the hysteresis in damping, in CNT reinforced polymer matrix and this is due to shear slip mechanism [9]. The energy dissipation in the CNT-polymer matrix nanocomposites occurs due to mismatch in elastic, dispersion of reinforcement and composition, interfacial shear-strength between CNT-polymer matrix. No research work was reported to find out the hysteresis in damping for aluminium based CNT reinforced nanocomposite material. Hence, this work aims to study the hysteresis damping behaviour of CNT reinforced aluminium and its alloys from various composition in weight (or) volume fraction, manufacturing techniques, dispersion pattern and temperature influence.

2. Materials design

The manufacturing methods used for developing Aluminium alloys-MWCNT and its microstructure observations are elaborated in this topic.

2.1 Fabrication of nanocomposites

AA 2024 (aluminium alloy) was used as a matrix material in a powder form of ~50 μm and reinforced with multi-walled carbon nanotubes (MWCNT) of 1 wt% with a purity of 95%. MWCNT used here were developed using catalysis of hydrocarbon. To obtain high purity and uniform dispersion of MWCNT, raw MWCNT with 95% purity undergone reflux with concentrated nitric acid at a temperature of 120°C for 10 h. After the process, by using distilled water MWCNT was continuously washed until the acidity gets dissolved. Then ethanol was added to influence the homogeneous dispersion. AA 2024 in powder form was mixed into MWCNT (ethanol solution) and the powders (AA 2024-CNT) distributed uniformly with ultrasonic shaker (mechanical stirring machine) for about 28 min. After the mechanical stirring process, mixed AA 2024 and MWCNT powders were kept in vacuum ~10^{-2} Pa at 120°C to remove the water content. The dried AA 2024 and MWCNT powders were smashed with ball milling process [10]. After fabricating AA 2024-MWCNT composite, to have a uniform dispersion cold isostatic pressing (CIP) along with extrusion process was used, proper structure of gas-atomized powders and to restrict the AA 2024-MWCNT formation [11]. Compared to other mechanical processes, CIP process shows a wide uniform distribution of density throughout the developed composite material, due to the distribution of pressure at isostatic state. The AA 2024-MWCNT (mixed powders) were packed in an Al bundle of 62 mm diameter into a rubberized bag and degasification process were done using vacuum pump mechanically operated at 0.01 Pa for about 20 min before

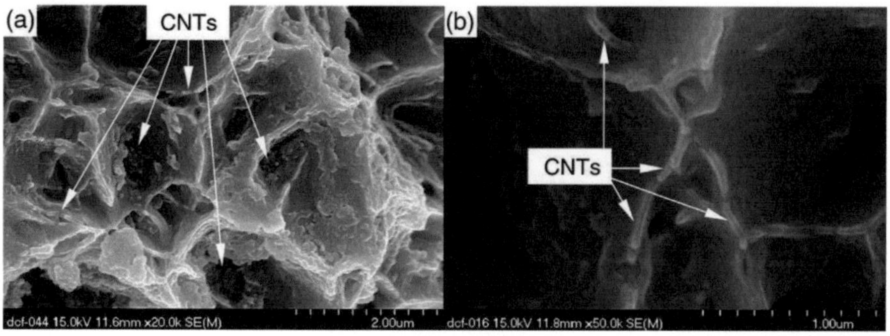

Figure 1.
Scanning electron microscope (SEM) image of (a) MWCNT reinforced in AA 2024 matrix and (b) magnified view of (a) [11, 12].

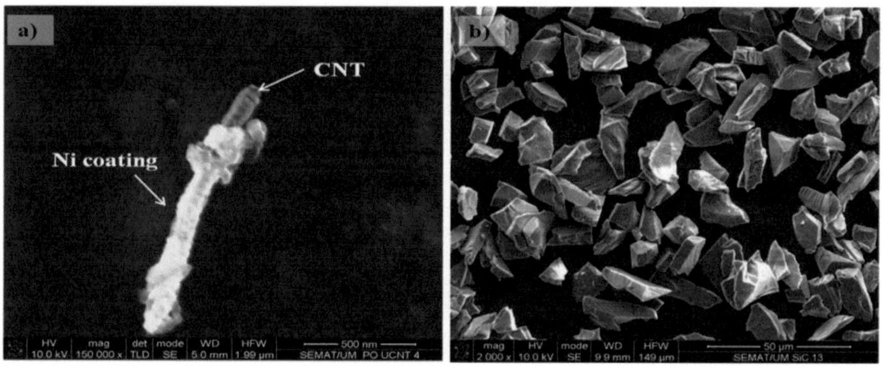

Figure 2.
(a) Nickel-coated MWCNT with an outer diameter 50 nm. (b) Silicon carbide (SiC) with a particle size of ~13 μm [13, 14].

the rubber gas gets wrapped. By using CIP process the mixed AA 2024-MWCNT powders were densified for 10 min at 300 MPa pressure [12]. After the CIP, using extrusion process composite billets were made at an extrusion ratio of 25:1 around 460°C of extrusion temperature. The SEM image of AA 2024-MWCNT composite shows the uniform dispersion of MWCNT reinforcement into the matrix material (**Figure 1**).

Aluminium silicon (AlSi) alloy powder-MWCNT composites [13] and AlSi-MWCNT-SiC are the hybrid composites developed to understand the damping behaviour of the material [14]. AlSi alloy powder with particle size of 325 mesh is used a matrix material.

Ni (Nickel) coated MWCNT (Cheaptubes, USA) was used as reinforcement with a purity of 95% (before coating), an outer diameter of 50 nm, length of 0.5–1.95 μm and Silicon Carbide (SiC) particles with a size approximately 13 μm [14]. To eliminate (or) restrict (or) control the formation of Al_4C_3 (brittle intermetallic) during the AlSi-MWCNT mixing, Ni coated MWCNT were used in this composite. The brittle intermetallic will reduce the damping property of the developed composites. The scanning electron microscope (SEM) images shows the Ni coated MWCNT and SiC particles [14] (**Figure 2**).

AlSi alloy powder and the corresponding reinforcements were mixed mechanically in a closed jar (stainless steel) with a milling balls of 10 mm diameter made of steel. The powder to ball ratio is of 10:1. The jar rotated at uniform speed around 40 rpm for 6 days in low energy ball milling [10, 13, 14].

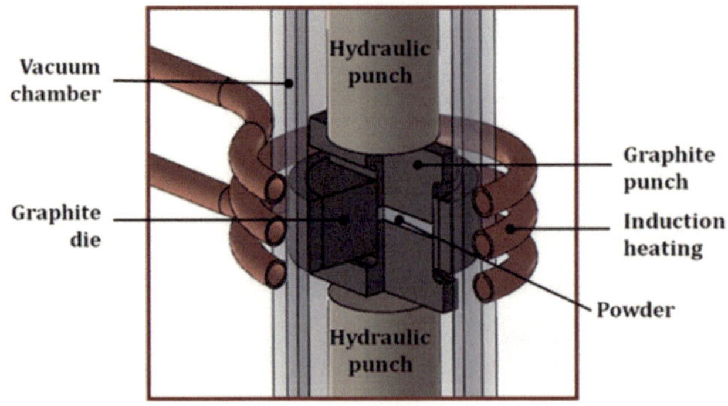

Figure 3.
Graphical representation of sintering process with hydraulic punch [14].

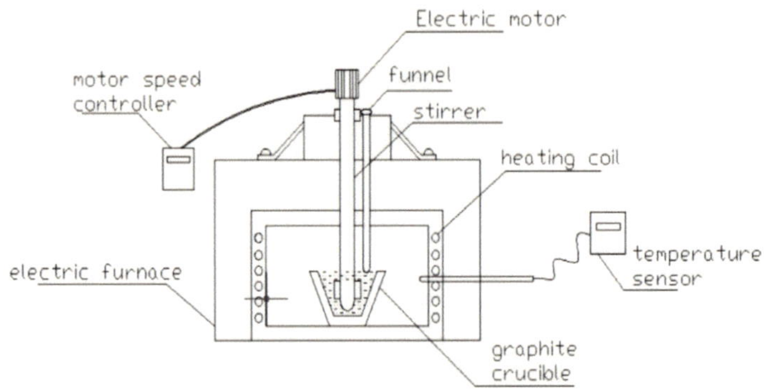

Figure 4.
Graphical representation of compo-casting process.

The mixed composite powder is divided and placed into the moulds made of graphite. AlSi alloy composites were developed using hot-pressing at a vacuum pressure of 10^{-6} Pa and heated using an induction furnace (high frequency). The developed composite mould then placed in the chamber, where the composite specimens will get compressed at 1 MPa and simultaneously heated to 500°C at a heating rate of 25°C/min. After the temperature reaches 500°C, pressure on the composite specimens is increased to 35 MPa without changing the heating rate. For the pressure of 35 MPa the temperature was steadily maintained for 550°C around 10 min [14]. Then the composite specimens are cooled until it reaches the room temperature (**Figure 3**).

AA 5083 material reinforced with MWCNT and developed into a composite by varying MWCNT composition of 1, 1.25, 1.5 and 1.75 in wt fraction using compo-casting (semi-solid state) method shown in **Figure 4** to improve damping and corrosion resistance property [15, 16]. To avoid the intermetallic bond between Al and MWCNT, compo-casting method was selected. A chemical bonding between Al and MWCNT will occur at 600°C [17]. The semi-solid-state temperature of Al was measured using thermocouple during casting process. Correspondingly, the MWCNT reinforcement was added into the matrix material and stirred for 2 min at a speed of 300 rpm. Then, molten melt was poured into the mould and desired composite material was developed [15, 16].

2.2 Microstructure observations of nanocomposites

The structure before and after milling of nanocomposites powder is presented which was obtained using scanning electron microscope (SEM) image, refer

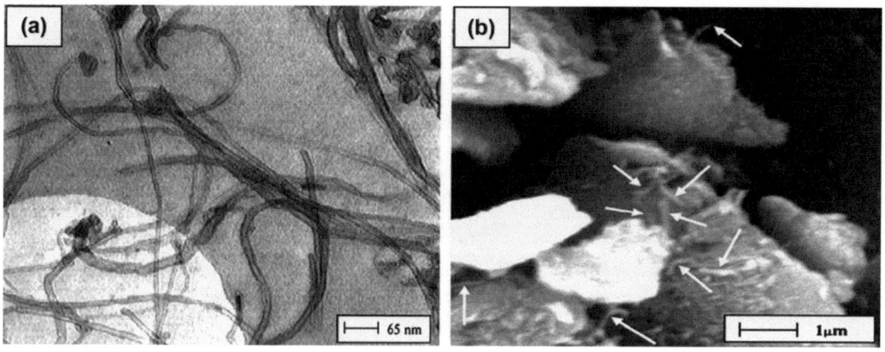

Figure 5.
(a) MWCNT view with high magnification and (b) dispersed MWCNT into the aluminium 2024 nanocomposite [11, 14, 18, 19].

Figure 6.
SEM image of aluminium 2024 matrix nanocomposite with uniform dispersion of MWCNT into the matrix [11, 18, 19].

Figure 5(a) and **(b)**. It shows that the MWCNT length have been varied due to the milling process. A nanocomposite is consisting of AA 2024 as matrix material in an average grain size of 30 nm and the corresponding reinforcement MWCNT with change in outer diameter and its length respectively was developed [18–21]. The SEM image shows the uniform dispersion of MWCNT into the AA 2024 matrix material, due to ball milling along with extrusion process shown in **Figure 5**. The SEM image of all composites in **Figure 6** shows the dispersion of CNT with less porosity, due to to hot pressing techniques. This shows the effectiveness of developing AlSi-MWCNT and AlSi-MWCNT-SiC composites using hot pressing techniques [14].

In **Figure 7(A)** AlSi-MWCNT composite image, the white zones in MWCNT is due to the Ni coated on MWCNT and in **Figure 7(B)** and **(C)** shows a cluster formation of MWCNT in AlSi-MWCNT-SiC composites [14]. Transmission Electron Microscope (TEM) analysis, Energy dispersive X-ray spectroscopy (EDX) and X-ray diffraction (XRD) were carried out to check the chemical bonding (brittle intermetallic reaction) between MWCNT and AlSi. No reaction was occurred due to the Ni coated MWCNT [12–22].

Figure 8 shows the TEM image of AlSI-MWCNT interfacial bonding established during hot pressing techniques. EDX analysis results shown in **Figure 9** is obtained from three areas such as AlSi, MWCNT and AlSI-MWCNT interference zone. The

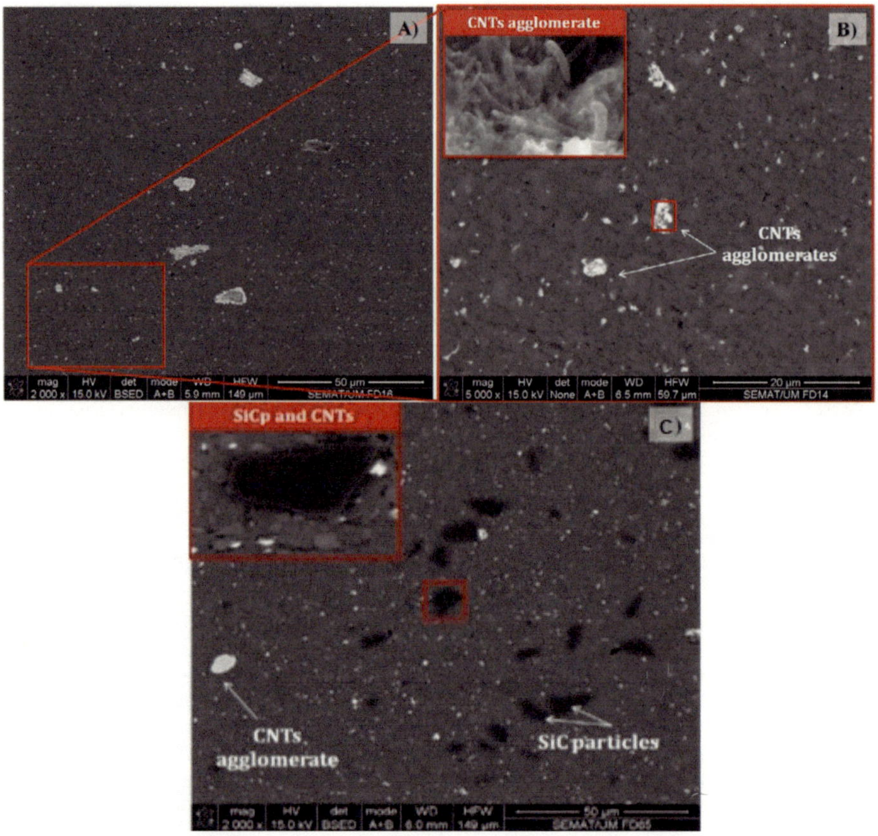

Figure 7.
The surface morphology image of (A) agglomeration of MWCNT, (B) MWCNT agglomeration with magnified view and (C) dispersion pattern of MWCNT and SiC into the AlSi matrix [13, 14].

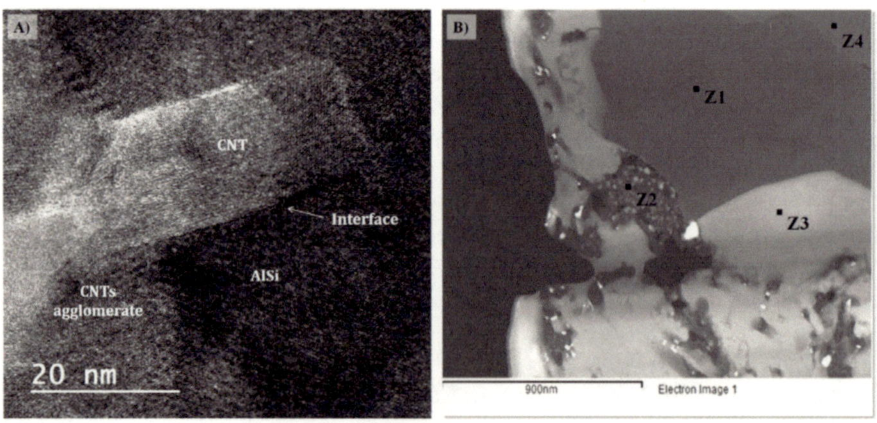

Figure 8.
Transmission electron microscope (TEM) image of (A) interface between Aluminium alloy–MWCNT (B) MWCNT agglomeration view with EDX [14, 23].

Z2 shows the presence of Ni and C (carbon), Z3 shows the presence of Al and Ni which indicate the intermetallic reaction compound between AlSi matrix material and MWCNT coated Ni reinforcement [14].

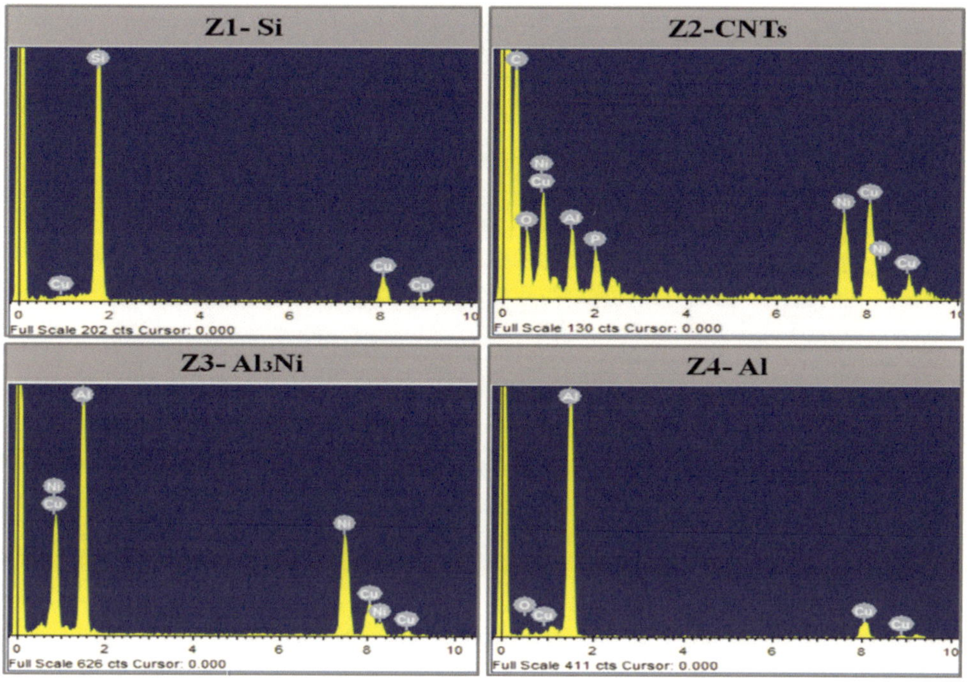

Figure 9.
EDX image of AlSi-reinforced MWCNT with interface and cluster formation [14].

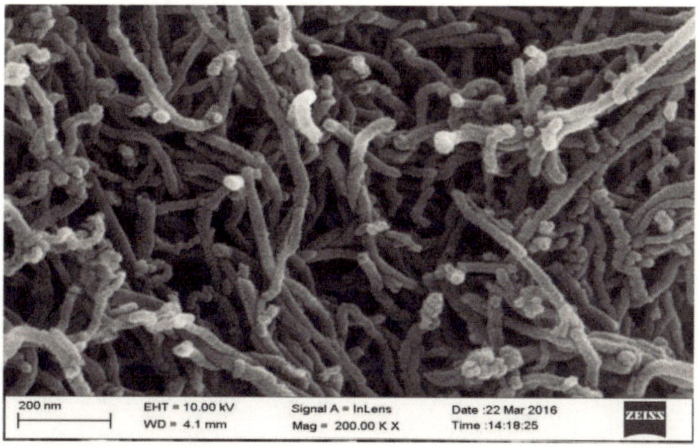

Figure 10.
Field emission scanning electron microscope (FESEM) image of closely bonded MWCNT [15, 16].

A uniform dispersion of AA 5083-MWCNT composites were attained using compo-casting method with minimum cluster formation. In **Figure 10**, field-emission scanning electron microscope (FESEM) image shows the structure MWCNT. **Figure 11** shows back-scattered electron (BSE) FESEM image of MWCNT reinforced into the matrix material. The addition of MWCNT into the matrix material is less than 2 wt fraction. Hence, the presence of nanotubes cannot be viewed in the XRD pattern. By using compo-casting method the brittle inter-metallic bonding can be neglected and improved damping property is achieved [16].

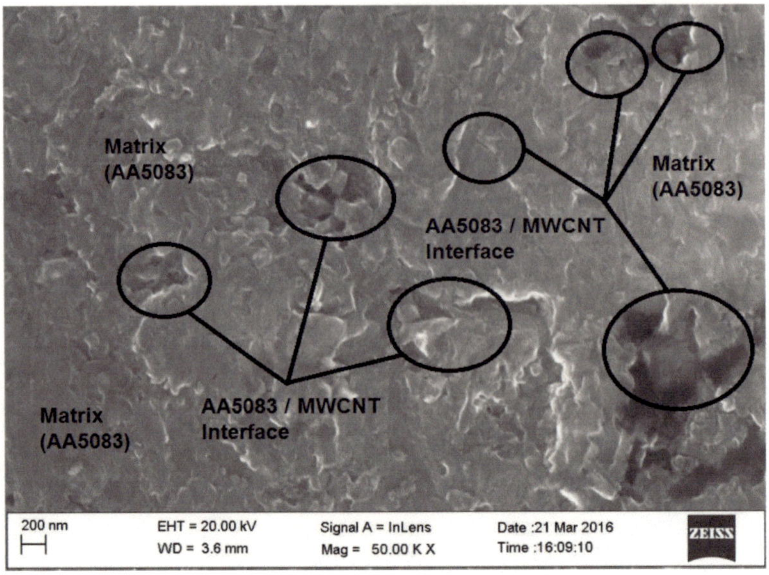

Figure 11.
Back-scattered electron (BSE) image of MWCNT interface and dispersion pattern [15].

3. Hysteresis damping study

3.1 Damping behaviour of AA 2024-MWCNT nanocomposites

The damping property of the developed AA 2024-MWCNT composites were experimentally tested using Dynamic Mechanical Analyser (DMA) [12–14]. Fixed as cantilever, and by varying the frequency to 0.5, 1, 5, 10 and 30 Hz at different temperatures (25–400°C). The damping properties are constant till the temperature is 199°C as shown in Tan delta curves of **Figure 12**, when the temperature is increased to 200°C

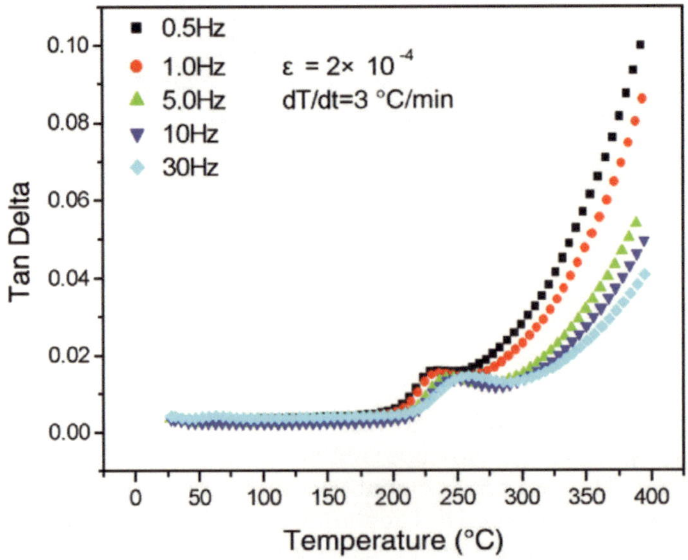

Figure 12.
Dynamic mechanical analyser (DMA) Tan delta vs. temperature curves of AA 2024-MWCNT composites for varying frequencies [12].

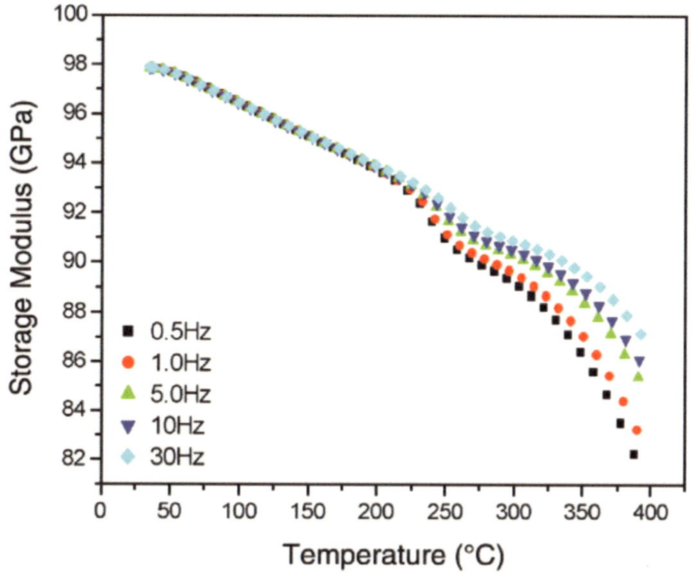

Figure 13.
Dynamic mechanical analyser (DMA) storage modulus vs. temperature curves of AA 2024-MWCNT composites for varying frequencies [12].

and above the changes in the damping property is intensely increasing at the lower frequency range [12]. Moreover, at 250°C there is a peak appears for 0.5, 1, 5, 10 and 30 Hz frequencies [24, 25]. Accordingly, increase in temperature shows increasing in the frequency [25]. At the temperature of 230°C, the damping property is more 10×10^{-3} for every single increase of frequency. At 0.5 Hz frequency and temperature of 400°C the damping property is 975×10^{-3}, which shows AA 2024 matrix reinforced MWCNT have a very high damping property at a maximum temperature [12, 24, 26, 27].

From **Figure 13**, it shows the storage modulus of DMA of the developed composites. It is observed that storage modulus of the AA 2024-MWCNT composites expressively reducing due to increase of temperature; simultaneously, the storage modulus is greater during higher frequency, when the temperature goes beyond 230°C [13, 24]. When the temperature reaches the maximum of 400°C [25], AA 2024-MWCNT composites shows some higher storage modulus around 82.3 GPa, than the AA 2024 alloy material of 71 GPa at the room temperature [24, 25]. Also, AA 2024-MWCNT shows some high mechanical properties without affecting the damping property of the material.

3.2 Damping behaviour of hybrid nanocomposites

The damping property of the developed composites AlSi-MWCNT and AlSi-MWCNT-SiC were experimentally tested using DMA, using temperature as a function [12–14] (room temperature to 300°C). Also, by varying frequencies to 1, 50 and 100 Hz, damping property of the composite material was measured [27, 28]. The damping property of the developed composites that were influenced by frequency and temperature is shown in **Figure 14**. At the initial condition of frequency (1 Hz), a steady change in increased damping property with high temperature is shown in **Figure 14(A)**. During the second frequency condition (50 Hz) comparable changes found at 150–200°C temperature shown in **Figure 14(B)**. At the final frequency value of 100 Hz, AlSi-MWCNT, AlSi-MWCNT-SiC found to decrease in their damping property due to maximum temperature as shown in **Figure 14(C)** [13, 24, 27, 28]. The curves shown in **Figure 14** indicate that damping property increases and decreases in a particular frequency level [28, 29].

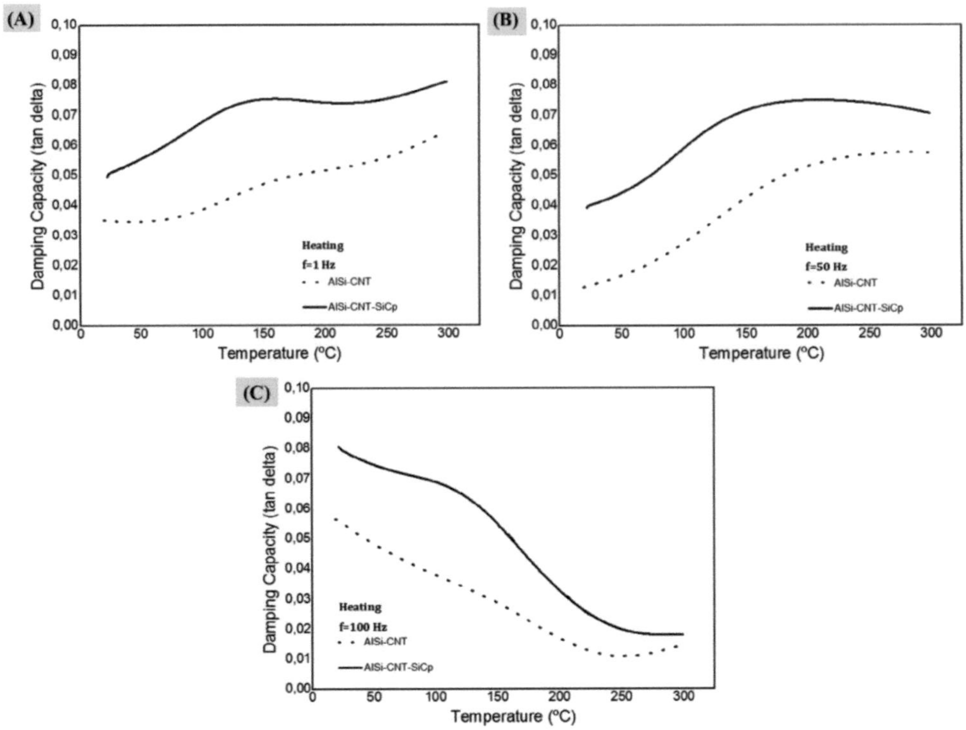

Figure 14.
Dynamic mechanical analyser (DMA) Tan delta vs. temperature curves of AlSi-MWCNT and AlSi-MWCNT-SiC composites for varying frequencies [13, 14].

Figure 14 shows Tan delta curves of AlSi-MWCNT, AlSi-MWCNT-SiC composites having maximum damping property value at the temperature of 150°C. At some point the damping property gets decreased due to MWCNT, SiC reinforcements experiencing low energy dissipation under maximum frequency condition [12–14, 29]. The cluster formation of MWCNT and SiC will affect the damping property of the material. Hence, the homogeneous dispersion of MWCNT into an AlSi matrix material increases the high interfacial-sliding and improves the damping property [29, 30].

AlSi-MWCNT, AlSI-MWCNT-SiC composites storage (or) dynamic modulus temperature function and its corresponding frequencies are shown in **Figure 15**. For all the frequency values, increase in temperature reduces the storage modulus and it can reduce the dynamic stiffness property [25, 31, 32]. Madeira et al. reported the same about the dynamic modulus in Al reinforced SiC particles [13].

In AlSi-MWCNT-SiC hybrid composites the storage modulus value is (~100 GPa) higher than AlSi-MWCNT composites (which is 73 GPa) in ambient (room) temperature shown in **Figure 15**. Based on the above results, adding MWCNT into the hybrid composites shows a very effective improvement in the modulus value [28, 31]. Therefore, Carbon nanotube provides a promising property which improves the damping property without compromising the mechanical properties. There are some of the major mechanisms involved in improving the damping properties are (i) interface sliding between MWCNT and AlSi matrix material (ii) a small amount of micro-voids due to ball milling method which dissipates energy are the two mechanisms which involves in improving the damping property of the developed material [24, 25, 32]. Based on the observations from Tan delta and storage modulus (or) dynamic young's modulus curves, it shows a hysteresis among the phases of heating and cooling [13, 25, 29].

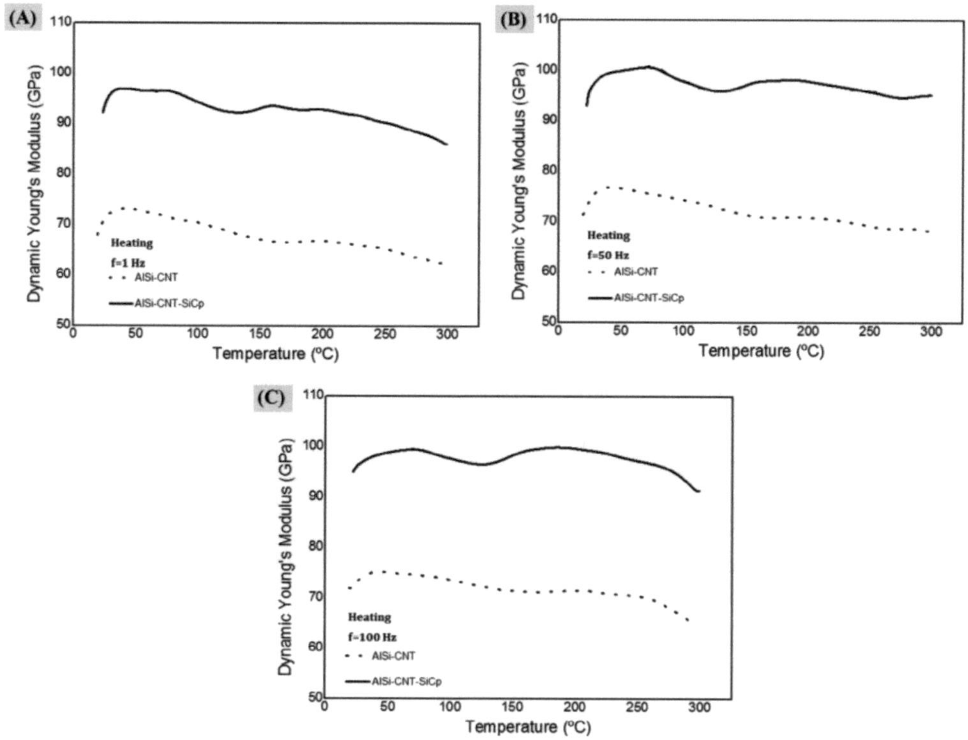

Figure 15.
Dynamic mechanical analyser (DMA) storage modulus vs. temperature curves of AlSi-MWCNT and AlSi-MWCNT-SiC composites for varying frequencies [13, 14].

4. Conclusions

From the above chapter, the following points are discussed:

1. The composites fabrication methods of Aluminium 2024 alloy, Aluminium Silicon alloy reinforced with MWCNT and SiC.

2. Dispersion pattern and interface between Aluminium alloys—MWCNT reinforcement into the matrix material observed with a minimum cluster formation is presented.

3. Ni coated MWCNT with varying composition into the matrix material are elaborated and Ni coated MWCNT resist the brittle intermetallic formation between matrix and the reinforcement.

4. Damping property was studied based on the dynamic mechanical analyser test.

5. Variation of frequencies and temperature shows the changes in damping property.

6. Hysteresis of damping for aluminium reinforced with MWCNT is explained.

7. Influence of MWCNT and damping mechanisms are briefed.

Notes/thanks/other declarations

Authors thank Kumaraguru College of Technology, Coimbatore, and Tamil Nadu – India for providing the required support to carry out the research work.

Acronyms and abbreviations

Al	aluminium
CNT	carbon nanotube
MWCNT	multi-wall carbon nanotube
AA	aluminium alloy
MMCs	metal matrix composites
AlSi	aluminium silicon alloy
SiC	silicon carbide
Al_4C_3	aluminium carbide
Ni	nickel
C	carbon
CIP	cold isostatic pressing
SEM	scanning electron microscope
FESEM	field-emission scanning electron microscope
TEM	transmission electron microscope
EDX	energy dispersive X-ray spectroscopy
X-RD	X-ray diffraction
BSE	back scattered electron
DMA	dynamic mechanical analyser

Author details

Paul Suresh Samuel Ratna Kumar and Savariar John Alexis*
Kumaraguru College of Technology, Coimbatore, Tamil Nadu, India

*Address all correspondence to: johnalexis@rediffmail.com

References

[1] Li G-c, Yue MA. Damping capacity of high strength-damping aluminum alloys prepared by rapid solidification and powder metallurgy process. Transaction of Nonferrous Metals Society China. 2012;**22**:1112-1117. DOI: 10.1016/S1003-6326(11)61291-0

[2] Ritchie IG, Pan ZL. High-damping metals and alloys. Metallurgical Transactions A. 1991;**22**:607-616. DOI: 10.1007/BF02670281

[3] Schaller R. Metal matrix composites, a smart choice for high damping materials. Journal of Alloys and Compounds. 2003;**355**:131-135. DOI: 10.1016/S0925-8388(03)00239-1

[4] Liu Q, Li M, Gu Y, Wang S, Zhang Y, Li Q. Interlocked CNT networks with high damping and storage modulus. Carbon. 2015;**86**:46-53. DOI: 10.1016/j.carbon.2015.01.014

[5] Khan SU, Li CY, Siddiqui NA, Kim JK. Vibration damping characteristics of carbon fiber-reinforced composites containing multi-walled carbon nanotubes. Composites Science and Technology. 2011;**71**(12):1486-1494. DOI: 10.1016/j.compscitech.2011.03.022

[6] Koratka NA, Wei B, Ajayan PM. Multifunctional structural reinforcement featuring carbon nanotube films. Composites Science and Technology. 2003;**63**(11):1525-1531. DOI: 10.1016/S0266-3538(03)00065-4

[7] Silva F. Fatigue on engine pistons–A compendium of case studies. Engineering Failure Analysis. 2006;**13**(3):480-492. DOI: 10.1016/j.engfailanal.2004.12.023

[8] Lin R, Lu C. Modeling of interfacial friction damping of carbon nanotube-based nanocomposites. Mechanical Systems and Signal Processing. 2010;**24**(8):2996-3012. DOI: 10.1016/j.ymssp.2010.06.003

[9] Zhang XC, Peng HX, Limmack AP, Scarpa F. Viscoelastic damping behaviour of cup stacked carbon nanotube modified epoxy nanocomposites with tailored interfacial condition and re-agglomeration. Composites Science and Technology. 2014;**105**:66-72. DOI: 10.1016/j.compscitech.2014.09.020

[10] Liu ZY, Xu SJ, Xiao BL. Effect of ball-milling time on mechanical properties of carbon nanotubes reinforced aluminium matrix composites. Composites Part A: Applied Science and Manufacturing. 2012;**43**(12):2161-2168. DOI: 10.1016/j.compositesa.2012.07.026

[11] Nouri N, Ziaei-Rad S. Fabrication and mechanical property prediction of carbon nanotube reinforced aluminum nanocomposites. Materials and Design. 2012;**34**:1-14. DOI: 10.1016/j.matdes.2011.07.047

[12] Deng CF, Wang DZ. Damping characteristics of carbon nanotube reinforced aluminum composite. Materials Letters. 2007;**61**:3229-3231. DOI: 10.1016/j.matlet.2006.11.073

[13] Madeira S, Miranda G. Study on damping capacity and dynamic Young's modulus of aluminium matrix composite reinforced with SiC particles. Ciência & Tecnologia dos Materiais. 2007;**29**:92-96. DOI: 10.1016/j.ctmat.2016.08.003

[14] Carvalho O, Miranda G. High temperature damping behavior and dynamic Young's modulus of AlSi–CNT–SiCp hybrid composite. Composite Structures. 2016;**141**:155-162. DOI: 10.1016/j.compstruct.2016.01.046

[15] Samuel Ratna Kumar PS, Robinson Smart DS, John Alexis S. Modal analysis of MWCNT reinforced AA5083 composite material. International

Journal of Civil Engineering and Technology. 2017;**8**(9):167-177. DOI: http://www.iaeme.com/IJCIET/issues.asp?JType=IJCIET&VType=8&IType=9

[16] Samuel Ratna Kumar PS, Robinson Smart DS, John Alexis S. Corrosion behaviour of aluminium metal matrix reinforced with multi-wall carbon nanotube. Journal of Asian Ceramic Societies. 2017;**5**(1):71-75. DOI: 10.1016/j.jascer.2017.01.004

[17] Chen B, Shen J. Solid-state interfacial reaction and load transfer efficiency in carbon nanotubes (CNTs)-reinforced aluminum matrix composites. Carbon. 2017;**114**(1):198-208. DOI: 10.1016/j.carbon.2016.12.013

[18] Esawi AMK. Effect of carbon nanotube (CNT) content on the mechanical properties of CNT-reinforced aluminium composites. Composites Science and Technology. 2010;**70**:2237-2241. DOI: 10.1016/j.compscitech.2010.05.004

[19] Wu Y, Kim GY. Carbon nanotube reinforced aluminum composite fabricated by semi-solid powder processing. Journal of Materials Processing Technology. 2011;**211**:1341-1347. DOI: 10.1016/j.jmatprotec.2011.03.007

[20] Yosida Y. High-temperature shrinkage of single-walled carbon nanotube bundles up to 1600 K. Journal of Applied Physics. 2000;**87**:3338-3341. DOI: 10.1063/1.372345

[21] Silvestre N, Faria B. Compressive behavior of CNT-reinforced aluminum composites using molecular dynamics. Composites Science and Technology. 2014;**90**:16-24. DOI: 10.1016/j.compscitech.2013.09.027

[22] Deng CF, Wang DZ, Zhang XX. Processing and properties of carbon nanotubes reinforced aluminum composites. Materials Science and Engineering: A. 2007;**444**(1-2):138-145. DOI: 10.1016/j.msea.2006.08057

[23] Lancin M, Marhic C. TEM study of carbon fibre reinforced aluminium matrix composites: Influence of brittle phases and interface on mechanical properties. Journal of the European Ceramic Society. 2000;**20**:1493-1503. DOI: 10.1016/S0955-2219(00)00021-2

[24] Liu G, Tang S, Ren W, Hu J, Li D. Damping peak and damping mechanism in $Al_{18}B_4O_{33}$w/Al composite containing Sn and Bi interfacial phases at room temperature. Materials and Design. 2013;**46**:916-921. DOI: 10.1016/j.matdes.2012.11.045

[25] Deng K-K, Li J-C, Nie K-B, Wang X-J, Fan J-F. High temperature damping behavior of as-deformed Mg matrix influenced by micron and submicron SiCp. Materials Science and Engineering: A. 2015;**624**:62-70. DOI: 10.1016/j.msea.2014.11.069

[26] Ajayan PM, Schadler LS, Giannaris C, Rubio A. Single-walled carbon nanotube–polymer composites: Strength and weakness. Advanced Materials. 2000;**12**(10):750-753. DOI: 10.1002/(SICI)1521-4095(200005)12:10<750::AID-ADMA750>3.0.CO;2-6

[27] Zhang Y, Ma N, Li X, Wang H. Study on damping capacity of aluminium composite reinforced with in situ $TiAl_3$ rod. Materials and Design. 2008;**29**(5):1057-1059. DOI: 10.1016/j.matdes.2007.04.001

[28] Gu J, Zhang X, Gu M, Gu M, Wang X. Internal friction peak and damping mechanism in high damping 6061Al/SiCp/gr hybrid metal matrix composite. Journal of Alloys and Compounds. 2004;**372**(1):304-308. DOI: 10.1016/j.jallcom.2003.10.021

[29] Formica G, Milicchio F, Lacarbonar W. Hysteretic damping

optimization in carbon nanotube nanocomposites. Composite Structures. 2018;**194**:633-642. DOI: 10.1016/j.compstruct.2018.04.027

[30] Besterci M. Preparation, microstructure and properties of Al–Al_4C_3 system produced by mechanical alloying. Materials and Design. 2006;**27**:416-421. DOI: 10.1016/j.matdes.2004.11.012

[31] Zhang J, Perez R, Lavernia E. Dislocation-induced damping in metal matrix composites. Journal of Materials Science. 1993;**28**(3):835-846. DOI: 10.1007/BF01151266

[32] Kang CS, Maeda K, Wang KJ, Wakasima K. Dynamic Young's modulus and internal friction in particulate SiC/Al composites. Acta Materialia. 1998;**46**(4):1209-1220. DOI: 10.1016/S1359-6454(97)00293-0

Section 3

Damage of Composites

Chapter 3

Matrix Cracking of Ceramic-Matrix Composites

Li Longbiao

Abstract

In this chapter, the matrix cracking of fiber-reinforced ceramic-matrix composites (CMCs) is investigated using the energy balance approach. The relationship between the matrix cracking stress, fiber and interface oxidation, and fiber failure is established. The effects of the fiber volume, interface shear stress and interface debonding energy, fiber failure, and oxidation temperature on the time-dependent matrix cracking stress are analyzed. The experimental matrix cracking stress of different fiber-reinforced CMCs is predicted using the present models.

Keywords: ceramic-matrix composites, matrix cracking, oxidation, interface shear stress

1. Introduction

Ceramic-matrix composites (CMCs) possess high specific strength and high specific modulus especially at elevated temperature and have already been applied in hot-section components in aeroengine [1]. However, at elevated temperature, the environment affects the mechanical performance of fiber-reinforced CMCs. The matrix cracking stress decreases with operation time due to the interface oxidation [2]. Many researchers performed investigations on matrix cracking of fiber-reinforced CMCs, i.e., the energy balance approach developed by Aveston et al. [3]; Budiansky et al. [4]; Rajan and Zok [5]; and Li [6, 7], and the fracture mechanics approach proposed by Marshall et al. [8], and McCartney [9]. At elevated temperature, the oxidative environment entered inside of fiber-reinforced CMCs through the microcrackings, and the lifetime is affected by the applied load and temperature [10, 11].

In this chapter, the effects of temperature, oxidation, and applied stress on the time-dependent matrix cracking stress of fiber-reinforced CMCs are investigated. The relationship between the matrix cracking stress, fiber and interface oxidation, and fiber failure is established. The effects of the fiber volume, interface shear stress and interface debonding energy, fiber failure, and oxidation temperature on the time-dependent matrix cracking stress are analyzed. The experimental matrix cracking stress of different fiber-reinforced CMCs are predicted.

2. Theoretical analysis

At elevated temperature, oxygen reacts with the interphase along the fiber length. The time-dependent interphase oxidation length can be determined using the following equation (11) [12].

IntechOpen

$$\zeta = \varphi_1 \left[1 - \exp\left(-\frac{\varphi_2 t}{b}\right)\right] \tag{1}$$

where φ_1 and φ_2 are parameters dependent on temperature and described using the Arrhenius type laws and b is a delay factor considering the deceleration of reduced oxygen activity.

The oxidation of fiber is assumed to be controlled by diffusion of oxygen gas through matrix cracks. By assuming that the fracture toughness of the fibers remains constant and that the fiber strength σ_0 is related to the mean oxidized layer thickness, the time dependence of the fiber strength can be determined by the following equation (12).

$$\sigma_0(t) = \sigma_0, t \leq \frac{1}{k}\left(\frac{K_{IC}}{Y\sigma_0}\right)^4 \tag{2}$$

$$\sigma_0(t) = \frac{K_{IC}}{Y\sqrt[4]{kt}}, t > \frac{1}{k}\left(\frac{K_{IC}}{Y\sigma_0}\right)^4 \tag{3}$$

When the fiber breaks, the stress carried by the intact and broken fibers can be determined using the global load sharing (GLS) criterion by Eq. (4) [13].

$$\frac{\sigma}{V_f} = T[1 - P(T)] + \langle T_b \rangle P(T) \tag{4}$$

where V_f denotes the fiber volume fraction, T denotes the load carried by intact fibers, $\langle T_b \rangle$ denotes the load carried by broken fibers, and $P(T)$ denotes the fiber failure probability.

$$P(T) = \eta P_a(T) + (1 - \eta) P_b(T) + P_c(T) \tag{5}$$

where η denotes the oxidation fiber fraction in the oxidation region and $P_a(T)$, $P_b(T)$, and $P_c(T)$ denote the fracture probability of oxidized fibers and unoxidized fibers in the oxidation region and interface debonding region, respectively.

2.1 Downstream stresses

Figure 1 shows the composite under loading of a remote uniform applied stress of σ with a long matrix cracking. A unit cell is extracted from the composite to analyze the microstress field of the damaged fiber-reinforced CMCs, as shown in **Figure 2**. The fiber and matrix axial stress in the interface oxidation and debonding region can be determined using the following equation.

$$\sigma_f^D(z) = \begin{cases} T - \frac{2\tau_f}{r_f} z, z \in (0, \zeta) \\ T - \frac{2\tau_f}{r_f}\zeta - \frac{2\tau_i}{r_f}(z - \zeta), z \in (\zeta, l_d) \end{cases} \tag{6}$$

$$\sigma_m^D(z) = \begin{cases} 2\frac{V_f}{V_m}\frac{\tau_f}{r_f} z, z \in (0, \zeta) \\ 2\frac{V_f}{V_m}\frac{\tau_f}{r_f}\zeta + 2\frac{V_f}{V_m}\frac{\tau_i}{r_f}(z - \zeta), z \in (\zeta, l_d) \end{cases} \tag{7}$$

In the fiber/matrix interface bonding region, the fiber and matrix axial stresses can be determined using the following equation.

$$\sigma_f^D = \frac{E_f}{E_c}\sigma \tag{8}$$

$$\sigma_m^D = \frac{E_m}{E_c}\sigma \tag{9}$$

2.2 Upstream stresses

The upstream region III as shown in **Figure 1** is so far away from the crack tip that the stress and strain fields are also uniform. The axial stress of the fiber and the matrix can be determined using the following equations.

$$\sigma_f^U = \frac{E_f}{E_c}\sigma \tag{10}$$

$$\sigma_m^U = \frac{E_m}{E_c}\sigma \tag{11}$$

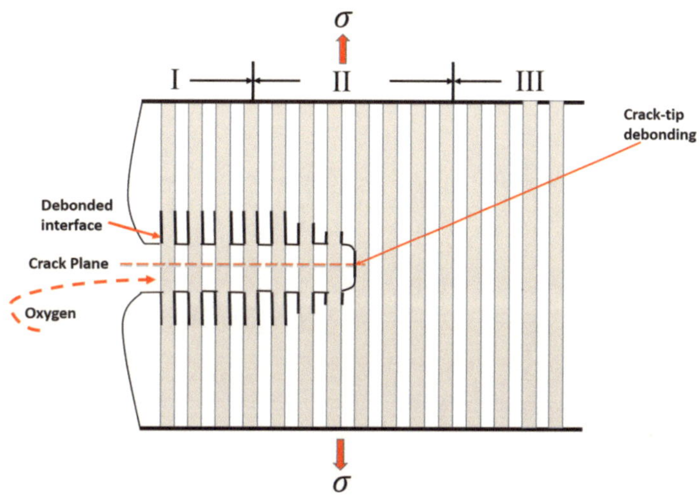

Figure 1.
The schematic of crack-tip, interface debonding and oxidation.

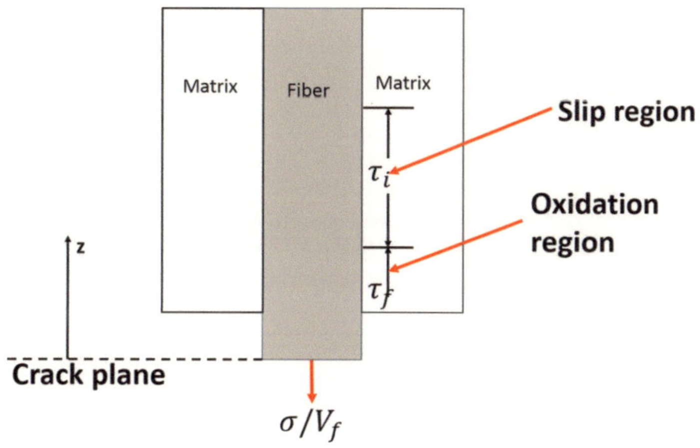

Figure 2.
The schematic of shear-lag model considering the interface oxidation and debonding.

2.3 Interface debonding

The fracture mechanics approach is used to determine the fiber/matrix interface debonding length [14].

$$\xi_{\mathrm{d}} = -\frac{F}{4\pi r_{\mathrm{f}}}\frac{\partial w_{\mathrm{f}}(0)}{\partial l_{\mathrm{d}}} - \frac{1}{2}\int_0^{l_{\mathrm{d}}} \tau_{\mathrm{i}} \frac{\partial v(z)}{\partial l_{\mathrm{d}}} dz \tag{12}$$

where $F(=\pi r_{\mathrm{f}}^2\sigma/V_{\mathrm{f}})$ denotes the fiber load at the matrix cracking plane, $w_{\mathrm{f}}(0)$ denotes the fiber axial displacement on the matrix cracking plane, and $v(z)$ denotes the relative displacement between the fiber and the matrix. Substituting $w_{\mathrm{f}}(z = 0)$ and $v(z)$ into Eq. (12), it leads to the form of Eq. (13).

$$\begin{aligned}&\frac{E_{\mathrm{c}}\tau_{\mathrm{i}}^2}{r_{\mathrm{f}}V_{\mathrm{m}}E_{\mathrm{f}}E_{\mathrm{m}}}(l_{\mathrm{d}}-\zeta)^2 + \frac{2E_{\mathrm{c}}\tau_{\mathrm{f}}\tau_{\mathrm{i}}}{r_{\mathrm{f}}V_{\mathrm{m}}E_{\mathrm{f}}E_{\mathrm{m}}}\zeta(l_{\mathrm{d}}-\zeta) - \frac{\sigma\tau_{\mathrm{i}}}{2V_{\mathrm{f}}E_{\mathrm{f}}}(l_{\mathrm{d}}-\zeta) - \frac{T\tau_{\mathrm{i}}}{2E_{\mathrm{f}}}(l_{\mathrm{d}}-\zeta)\\&+\frac{r_{\mathrm{f}}\sigma T}{4V_{\mathrm{f}}E_{\mathrm{f}}} - \frac{r_{\mathrm{f}}\sigma^2}{4V_{\mathrm{f}}E_{\mathrm{c}}} - \frac{\tau_{\mathrm{f}}\sigma\zeta}{2V_{\mathrm{f}}E_{\mathrm{f}}} - \frac{\tau_{\mathrm{f}}T}{2E_{\mathrm{f}}}\zeta + \frac{E_{\mathrm{c}}\tau_{\mathrm{f}}^2}{r_{\mathrm{f}}V_{\mathrm{m}}E_{\mathrm{f}}E_{\mathrm{m}}}\zeta^2 - \xi_{\mathrm{d}} = 0\end{aligned} \tag{13}$$

Solving Eq. (13), the fiber/matrix interface debonding length is determined by Eq. (14).

$$\begin{aligned}l_{\mathrm{d}} = {}&\left(1-\frac{\tau_{\mathrm{f}}}{\tau_{\mathrm{i}}}\right)\zeta + \frac{r_{\mathrm{f}}V_{\mathrm{m}}E_{\mathrm{m}}}{4E_{\mathrm{c}}\tau_{\mathrm{i}}}\left(\frac{\sigma}{V_{\mathrm{f}}}+T\right) - \left\{\frac{r_{\mathrm{f}}^2V_{\mathrm{m}}E_{\mathrm{m}}T^2}{4E_{\mathrm{c}}\tau_{\mathrm{i}}^2}\left[\frac{V_{\mathrm{m}}E_{\mathrm{m}}}{4E_{\mathrm{c}}}\left(\frac{\sigma}{V_{\mathrm{f}}T}+1\right)^2\right.\right.\\&\left.\left.+\frac{V_{\mathrm{f}}E_{\mathrm{f}}}{E_{\mathrm{c}}}\left(\frac{\sigma}{V_{\mathrm{f}}T}\right)^2 - \frac{\sigma}{V_{\mathrm{f}}T}\right] + \frac{r_{\mathrm{f}}V_{\mathrm{m}}E_{\mathrm{f}}E_{\mathrm{m}}}{E_{\mathrm{c}}\tau_{\mathrm{i}}^2}\xi_{\mathrm{d}}\right\}^{\frac{1}{2}}\end{aligned} \tag{14}$$

2.4 Matrix cracking stress

The energy relationship to evaluate the steady-state matrix cracking stress is determined by Eq. (15) [4].

$$\begin{aligned}&\frac{1}{2}\int_{-\infty}^{\infty}\left[\frac{V_{\mathrm{f}}}{E_{\mathrm{f}}}\left(\sigma_{\mathrm{f}}^{\mathrm{U}}-\sigma_{\mathrm{f}}^{\mathrm{D}}\right)^2 + \frac{V_{\mathrm{m}}}{E_{\mathrm{m}}}\left(\sigma_{\mathrm{m}}^{\mathrm{U}}-\sigma_{\mathrm{m}}^{\mathrm{D}}\right)^2\right]dz + \frac{1}{2\pi R^2G_{\mathrm{m}}}\int_{-l_{\mathrm{d}}}^{l_{\mathrm{d}}}\int_{r_{\mathrm{f}}}^{R}\left(\frac{r_{\mathrm{f}}\tau_{\mathrm{i}}(z)}{r}\right)2\pi r dr dz\\&= V_{\mathrm{m}}\xi_{\mathrm{m}} + \left(\frac{4V_{\mathrm{f}}l_{\mathrm{d}}}{r_{\mathrm{f}}}\right)\xi_{\mathrm{d}}\end{aligned} \tag{15}$$

where ξ_{m} is the matrix fracture energy and G_{m} is the matrix shear modulus. Substituting the fiber and matrix axial stresses in Eqs. (6)–(11) and the fiber/matrix interface debonding length of Eq. (14) into Eq. (15), the energy balance equation leads to the form of the following equation.

$$\eta_1\sigma^2 + \eta_2\sigma + \eta_3 = 0 \tag{16}$$

where

$$\eta_1 = \frac{l_{\mathrm{d}}}{E_{\mathrm{c}}} \tag{17}$$

$$\eta_2 = -\frac{2V_f T l_d}{E_c} \tag{18}$$

$$\eta_3 = \frac{V_f l_d T^2}{E_f} - \frac{V_f}{E_f}\left(\frac{2\tau_f}{r_f}\right)\zeta(2l_d - \zeta)T - \frac{V_f}{E_f}\left(\frac{2\tau_i}{r_f}\right)(l_d - \zeta)^2 T$$
$$+\frac{4}{3}\left(\frac{V_f E_c}{V_m E_f E_m}\right)\left(\frac{\tau_i}{r_f}\right)^2\left[(l_d - \zeta)^3 + \left(\frac{\tau_f}{\tau_i}\right)^2\zeta^3\right] + \left(\frac{4V_f E_c}{V_m E_f E_m}\right)\frac{\tau_f \tau_i}{r_f^2}\zeta(l_d - \zeta)$$
$$\times\left[l_d - \left(1 - \frac{\tau_f}{\tau_i}\right)\zeta\right] - V_m \xi_m - \left(\frac{4V_f l_d}{r_f}\right)\xi_d \tag{19}$$

3. Result and discussion

The effects of fiber volume, fiber/matrix interface debonding energy, interface shear stress, fiber strength, and oxidation temperature on the matrix cracking stress, interface oxidation, and interface debonding are analyzed.

3.1 Effect of the fiber volume on matrix cracking stress

The matrix cracking stress, fiber/matrix interface debonding length, and fiber/matrix interface oxidation length versus the oxidation time curves corresponding to different fiber volume of V_f = 30% and 35% are shown in **Figure 3**. When the fiber volume increases, the matrix cracking stress and the fiber/matrix interface oxidation length increase, and the fiber/matrix interface debonding length decreases.

When the fiber volume is V_f = 30%, the matrix cracking stress decreases from σ_{mc} = 86 to 44 MPa after t = 10 h oxidation at elevated temperature of T_{em} = 800°C; the fiber/matrix interface debonding length first decreases from l_d/r_f = 6.9 to 6.7 after t = 2 h oxidation at elevated temperature of T_{em} = 800°C and then increases to l_d/r_f = 9.0 after t = 10 h oxidation at elevated temperature of T_{em} = 800°C, and the fiber/matrix interface oxidation length increases from ζ/l_d = 0 to 0.94 after t = 10 h oxidation at elevated temperature of T_{em} = 800°C.

When the fiber volume is V_f = 35%, the matrix cracking stress decreases from σ_{mc} = 97 to 50 MPa after t = 10 h oxidation at elevated temperature of T_{em} = 800°C; the fiber/matrix interface debonding length first decreases from l_d/r_f = 5.8 to 5.7 after t = 1.7 h oxidation at elevated temperature of T_{em} = 800°C and then increases to l_d/r_f = 8.4 after t = 10 h oxidation at elevated temperature of T_{em} = 800°C, and the fiber/matrix interface oxidation length increases from ζ/l_d = 0 to 1.0 after t = 10 h oxidation at elevated temperature of T_{em} = 800°C.

3.2 Effect of the fiber/matrix interface debonding energy on matrix cracking stress

The matrix cracking stress, fiber/matrix interface debonding length, and fiber/matrix interface oxidation length versus the oxidation time curves corresponding to different interface debonding energies of ξ_d/ξ_m = 0.1 and 0.2 are shown in **Figure 4**. When the fiber/matrix interface debonding energy increases, the matrix cracking stress and the interface oxidation length increase, and the interface debonding length decreases.

When the fiber/matrix interface debonding energy is ξ_d/ξ_m = 0.1, the matrix cracking stress decreases from σ_{mc} = 86 MPa to 32 MPa after t = 10 h oxidation at elevated temperature of T_{em} = 800°C; the fiber/matrix interface debonding length

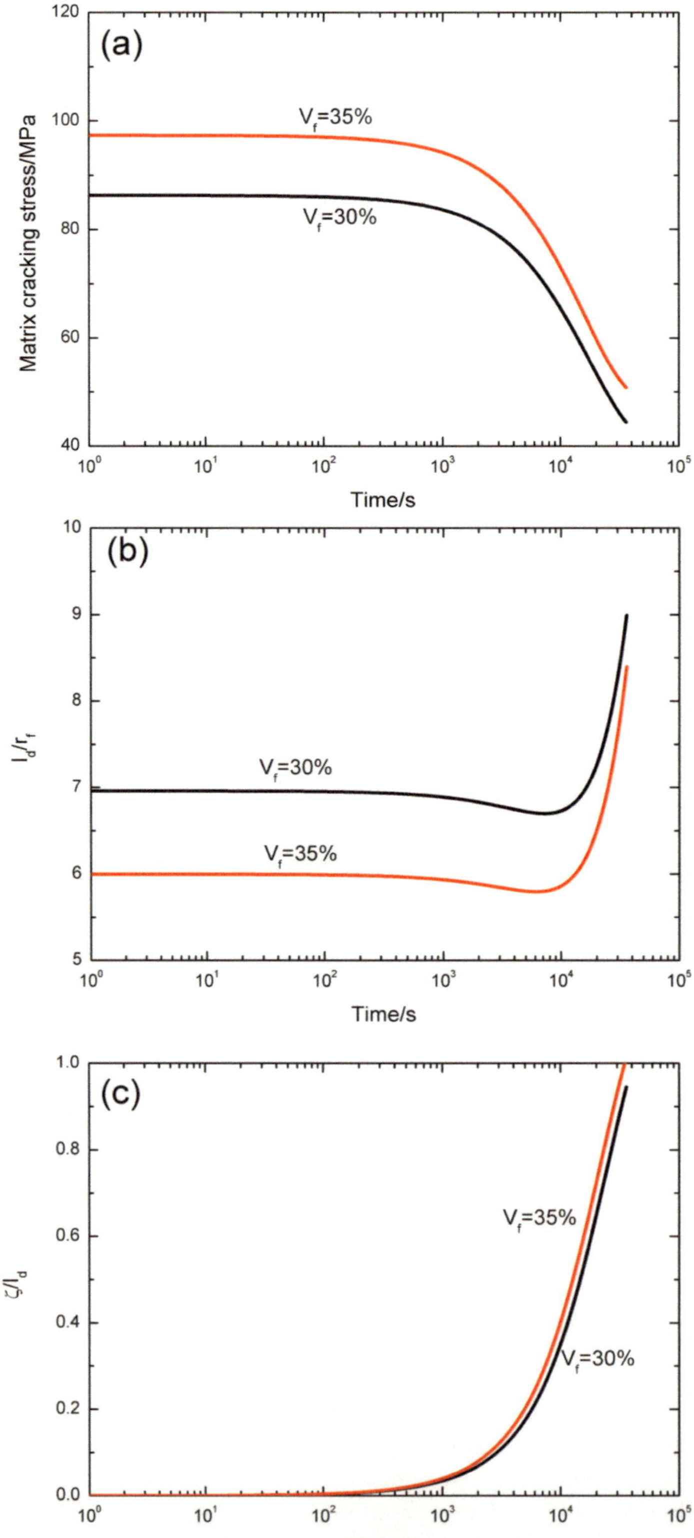

Figure 3.
(a) The matrix cracking stress versus the oxidation time, (b) the fiber/matrix interface debonding length versus the oxidation time, and (c) the fiber/matrix interface oxidation length versus the oxidation time corresponding to different fiber volume of V_f = 30 and 35%.

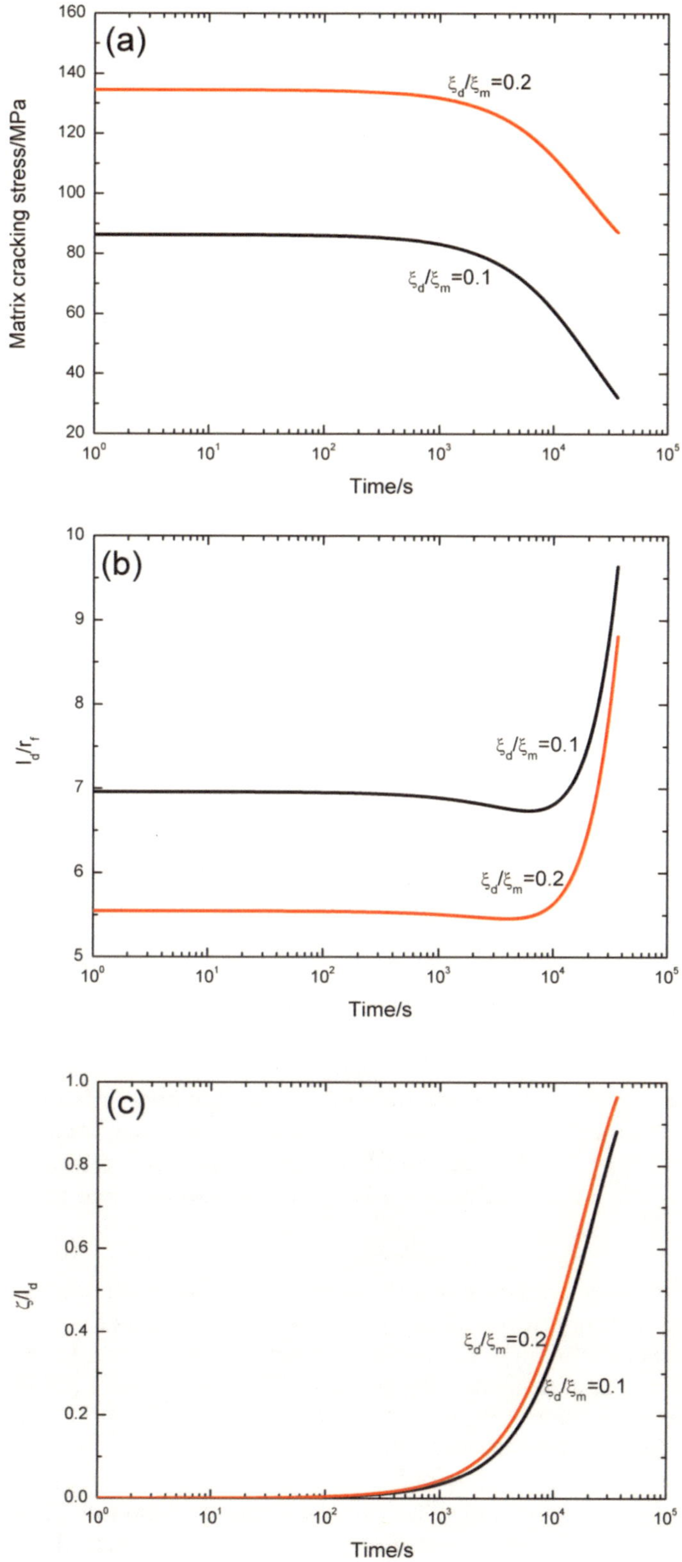

Figure 4.
(a) The matrix cracking stress versus the oxidation time, (b) the fiber/matrix interface debonding length versus the oxidation time, and (c) the fiber/matrix interface oxidation length versus the oxidation time corresponding to different interface debonding energy of ξ_d/ξ_m = 0.1 and 0.2.

first decreases from l_d/r_f = 6.9 to 6.7 after t = 1.7 h oxidation at elevated temperature of T_{em} = 800°C and then increases to l_d/r_f = 9.6 after t = 10 h oxidation at elevated temperature of T_{em} = 800°C, and the fiber/matrix interface oxidation length increases from ζ/l_d = 0 to 0.88 after t = 10 h oxidation at elevated temperature of T_{em} = 800°C.

When the fiber/matrix interface debonding energy is ξ_d/ξ_m = 0.2, the matrix cracking stress decreases from σ_{mc} = 134 to 87 MPa after t = 10 h oxidation at elevated temperature of T_{em} = 800°C; the fiber/matrix interface debonding length first decreases from l_d/r_f = 5.5 to 5.4 after t = 1.1 h oxidation at elevated temperature of T_{em} = 800°C and then increases to l_d/r_f = 8.8 after t = 10 h oxidation at elevated temperature of T_{em} = 800°C, and the fiber/matrix interface oxidation length increases from ζ/l_d = 0 to 0.96 after t = 10 h oxidation at elevated temperature of T_{em} = 800°C.

3.3 Effect of the fiber/matrix interface shear stress on the matrix cracking stress

The matrix cracking stress, fiber/matrix interface debonding length, and the fiber/matrix interface oxidation length versus the oxidation time curves corresponding to different fiber/matrix interface shear stress of τ_i = 20 and 10 MPa are shown in **Figure 5**. When the fiber/matrix interface shear stress in the slip region increases, the matrix cracking stress and the fiber/matrix interface oxidation length increase, and the fiber/matrix interface debonding length decreases.

When the fiber/matrix interface shear stress is τ_i = 10 MPa, the matrix cracking stress decreases from σ_{mc} = 67 to 43 MPa after t = 10 h oxidation at elevated temperature of T_{em} = 800°C; the fiber/matrix interface debonding length first decreases from l_d/r_f = 8.0 to 7.7 after t = 2.6 h oxidation at elevated temperature of T_{em} = 800°C and then increases to l_d/r_f = 9.1 after t = 10 h oxidation at elevated temperature of T_{em} = 800°C, and the fiber/matrix interface oxidation length increases from ζ/l_d = 0 to 0.93 after t = 10 h oxidation at elevated temperature of T_{em} = 800°C.

When the fiber/matrix interface shear stress is τ_i = 20 MPa, the matrix cracking stress decreases from σ_{mc} = 102 to 45 MPa after t = 10 h oxidation at elevated temperature of T_{em} = 800°C; the fiber/matrix interface debonding length first decreases from l_d/r_f = 6.2 to 6.0 after t = 1.7 h oxidation at elevated temperature of T_{em} = 800°C and then increases to l_d/r_f = 8.9 after t = 10 h oxidation at elevated temperature of T_{em} = 800°C, and the fiber/matrix interface oxidation length increases from ζ/l_d = 0 to 0.95 after t = 10 h oxidation at elevated temperature of T_{em} = 800°C.

The matrix cracking stress, fiber/matrix interface debonding length, and the fiber/matrix interface oxidation length versus the oxidation time curves corresponding to different fiber/matrix interface shear stress of τ_f = 1 and 5 MPa are shown in **Figure 6**. When the fiber/matrix interface shear stress in the oxidation region increases, the matrix cracking stress and the fiber/matrix interface oxidation length increase, and the fiber/matrix interface debonding length decreases.

When the fiber/matrix interface shear stress is τ_f = 1 MPa, the matrix cracking stress decreases from σ_{mc} = 86 to 28 MPa after t = 10 h oxidation at elevated temperature of T_{em} = 800°C; the fiber/matrix interface debonding length first decreases from l_d/r_f = 6.9 to 6.7 after t = 1.6 h oxidation at elevated temperature of T_{em} = 800°C and then increases to l_d/r_f = 9.8 after t = 10 h oxidation at elevated temperature of T_{em} = 800°C, and the fiber/matrix interface oxidation length increases from ζ/l_d = 0 to 0.86 after t = 10 h oxidation at elevated temperature of T_{em} = 800°C.

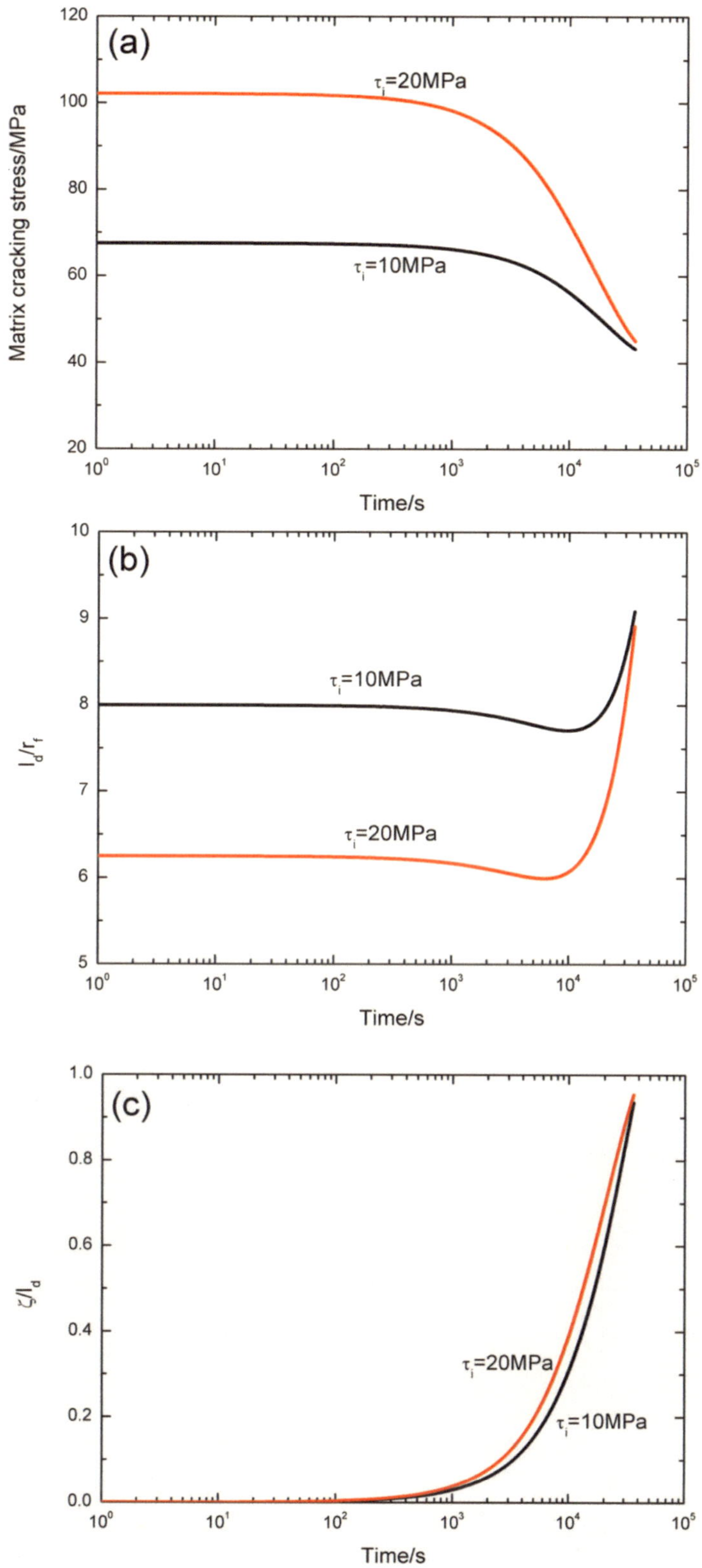

Figure 5.
(a) The matrix cracking stress versus the oxidation time, (b) the fiber/matrix interface debonding length versus the oxidation time, and (c) the fiber/matrix interface oxidation length versus the oxidation time corresponding to different interface shear stress of τ_i = 10 and 20 MPa.

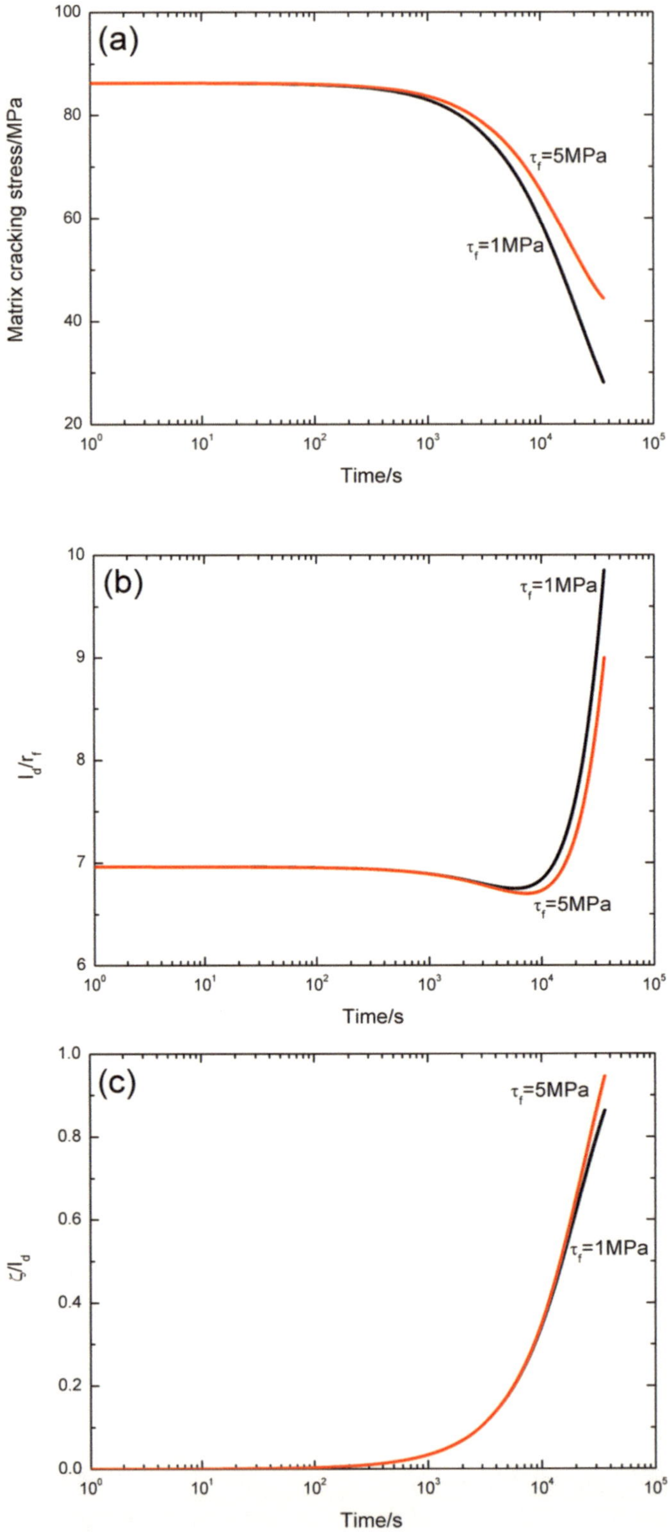

Figure 6.
(a) The matrix cracking stress versus the oxidation time, (b) the fiber/matrix interface debonding length versus oxidation time, and (c) the fiber/matrix interface oxidation length versus the oxidation time corresponding to different interface shear stress of τ_f = 1 and 5 MPa.

When the fiber/matrix interface shear stress is τ_f = 5 MPa, the matrix cracking stress decreases from σ_{mc} = 86 to 44 MPa after t = 10 h oxidation at elevated temperature of T_{em} = 800°C; the fiber/matrix interface debonding length first decreases from l_d/r_f = 6.9 to 6.7 after t = 2 h oxidation at elevated temperature of T_{em} = 800°C and then increases to l_d/r_f = 9 after t = 10 h oxidation at elevated temperature of T_{em} = 800°C, and the fiber/matrix interface oxidation length increases from ζ/l_d = 0 to 0.94 after t = 10 h oxidation at elevated temperature of T_{em} = 800°C.

3.4 Effect of the fiber strength on matrix cracking stress

The matrix cracking stress, fiber/matrix interface debonding length, and fiber/matrix interface oxidation length versus the oxidation time curves corresponding to different fiber strength of σ_0 = 1 and 2 GPa are shown in **Figure 7**. When the fiber strength increases, the matrix cracking stress and the fiber/matrix interface debonding length increase, and the fiber/matrix interface oxidation length decreases.

When the fiber strength is σ_0 = 1 GPa, the matrix cracking stress decreases from σ_{mc} = 38 to 18 MPa after t = 10 h oxidation at elevated temperature of T_{em} = 800°C; the fiber/matrix interface debonding length first decreases from l_d/r_f = 14.9 to 14.7 after t = 2 h oxidation at elevated temperature of T_{em} = 800°C and then increases to l_d/r_f = 16.3 after t = 10 h oxidation at elevated temperature of T_{em} = 800°C, and the fiber/matrix interface oxidation length increases from ζ/l_d = 0 to 0.52 after t = 10 h oxidation at elevated temperature of T_{em} = 800°C.

When the fiber strength is σ_0 = 2 GPa, the matrix cracking stress decreases from σ_{mc} = 43 to 21 MPa after t = 10 h oxidation at elevated temperature of T_{em} = 800°C; the fiber/matrix interface debonding length first decreases from l_d/r_f = 15.3 to 15.2 after t = 2.3 h oxidation at elevated temperature of T_{em} = 800°C and then increases to l_d/r_f = 16.6 after t = 10 h oxidation at elevated temperature of T_{em} = 800°C, and the fiber/matrix interface oxidation length increases from ζ/l_d = 0 to 0.51 after t = 10 h oxidation at elevated temperature of T_{em} = 800°C.

3.5 Effect of the temperature on matrix cracking stress

The matrix cracking stress, fiber/matrix interface debonding length, and fiber/matrix interface oxidation length versus the oxidation time curves corresponding to different oxidation temperature of T_{em} = 600 and 800°C are shown in **Figure 8**. When the oxidation temperature increases, the matrix cracking stress decreases, and the fiber/matrix interface oxidation length and the interface debonding length increase.

When the oxidation temperature is T_{em} = 600°C, the matrix cracking stress decreases from σ_{mc} = 86.3 to 73 MPa after t = 10 h oxidation; the fiber/matrix interface debonding length decreases from l_d/r_f = 6.9 to 6.7 after t = 10 h oxidation, and the fiber/matrix interface oxidation length increases from ζ/l_d = 0 to 0.19 after t = 10 h oxidation.

When the oxidation temperature is T_{em} = 800°C, the matrix cracking stress decreases from σ_{mc} = 86 to 44 MPa after t = 10 h oxidation; the fiber/matrix interface debonding length first decreases from l_d/r_f = 6.9 to 6.7 after t = 2 h oxidation and then increases to l_d/r_f = 9 after t = 10 h oxidation, and the fiber/matrix interface oxidation length increases from ζ/l_d = 0 to 0.94 after t = 10 h oxidation.

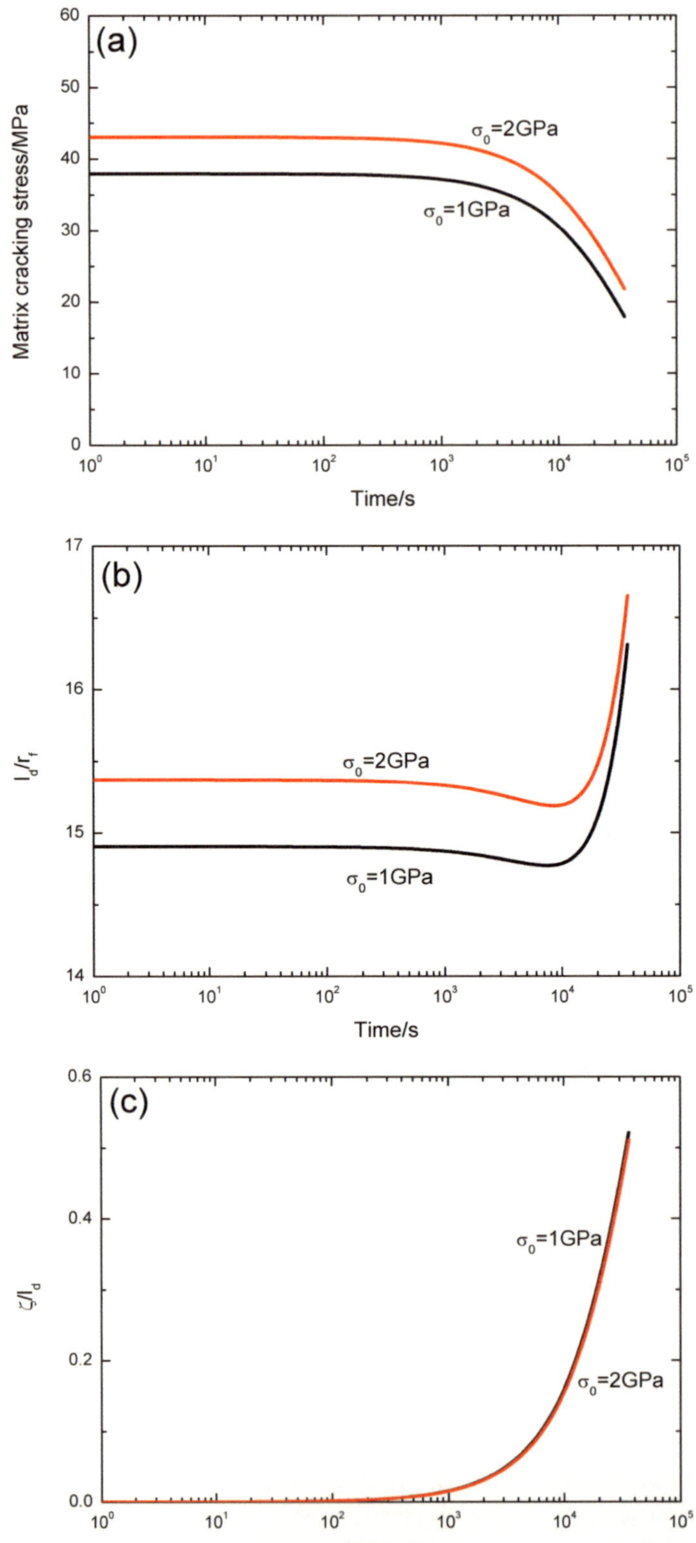

Figure 7.
(a) The matrix cracking stress versus the oxidation time, (b) the fiber/matrix interface debonding length versus the oxidation time, and (c) the fiber/matrix interface oxidation length versus the oxidation time corresponding to different fiber strength of σ_0 = *1 and 2 GPa.*

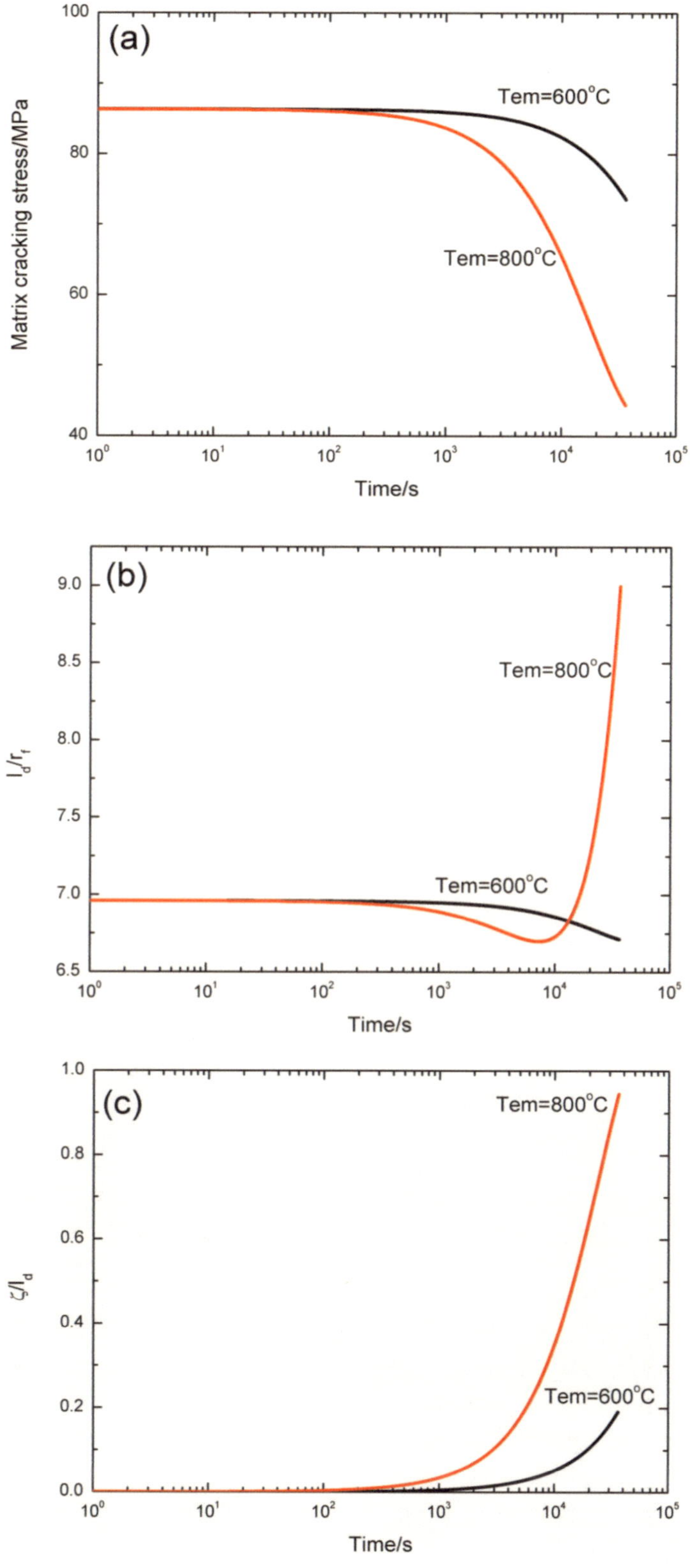

Figure 8.
(a) The matrix cracking stress versus the oxidation time, (b) the fiber/matrix interface debonding length versus the oxidation time, and (c) the fiber/matrix interface oxidation length versus the oxidation time corresponding to different oxidation temperature of T_{em} = 600 and 800°C.

4. Experimental comparison

Barsoum et al. [15] investigated the matrix crack initiation in fiber-reinforced CMCs. The experimental and predicted matrix cracking stress versus the fiber volume corresponding to different fiber/matrix interface debonding energy of SiC/borosilicate, SiC/LAS, and C/borosilicate composites is shown in **Figures 9–11**.

For the SiC/borosilicate composite, the predicted matrix cracking stress with the fiber/matrix interface debonding energy range of ξ_d/ξ_m = 0.05, 0.1, and 0.2 agrees with the experimental data corresponding to the fiber volume changing from 10 to 50%, as shown in **Figure 9**. When the fiber/matrix interface debonding energy is ξ_d/ξ_m = 0.05, the matrix cracking stress increases from σ_{mc} = 49.4 MPa at the fiber volume of V_f = 10% to σ_{mc} = 334 MPa at the fiber volume of V_f = 50%; when the fiber/matrix interface debonding energy is ξ_d/ξ_m = 0.1, the matrix cracking stress increases from σ_{mc} = 52.2 MPa at the fiber volume of V_f = 10% to σ_{mc} = 371 MPa at the fiber volume V_f = 50%; and when the fiber/matrix interface debonding energy is ξ_d/ξ_m = 0.2, the matrix cracking stress increases from σ_{mc} = 57.1 MPa at the fiber volume of V_f = 10% to σ_{mc} = 428 MPa at the fiber volume of V_f = 50%.

For the SiC/LAS composite, the matrix cracking stress with the interface debonding energy range of ξ_d/ξ_m = 0.01, 0.02, and 0.03 agrees with the experimental data corresponding to the fiber volume changing from 30 to 50%, as shown in **Figure 10**. When the fiber/matrix interface debonding energy is ξ_d/ξ_m = 0.01, the matrix cracking stress increases from σ_{mc} = 228 MPa at the fiber volume of V_f = 30% to σ_{mc} = 423 MPa at the fiber volume of V_f = 50%; when the fiber/matrix interface debonding energy is ξ_d/ξ_m = 0.02, the matrix cracking stress increases from σ_{mc} = 257 MPa at the fiber volume of V_f = 30% to σ_{mc} = 489 MPa at the fiber volume of V_f = 50%; and when the fiber/matrix interface debonding energy is ξ_d/ξ_m = 0.03, the matrix cracking stress increases from σ_{mc} = 281 MPa at the fiber volume of V_f = 30% to σ_{mc} = 542 MPa at the fiber volume of V_f = 50%.

For the C/borosilicate composite, the predicted matrix cracking stress with the fiber/matrix interface debonding energy range of ξ_d/ξ_m = 0.01, 0.02, and 0.03

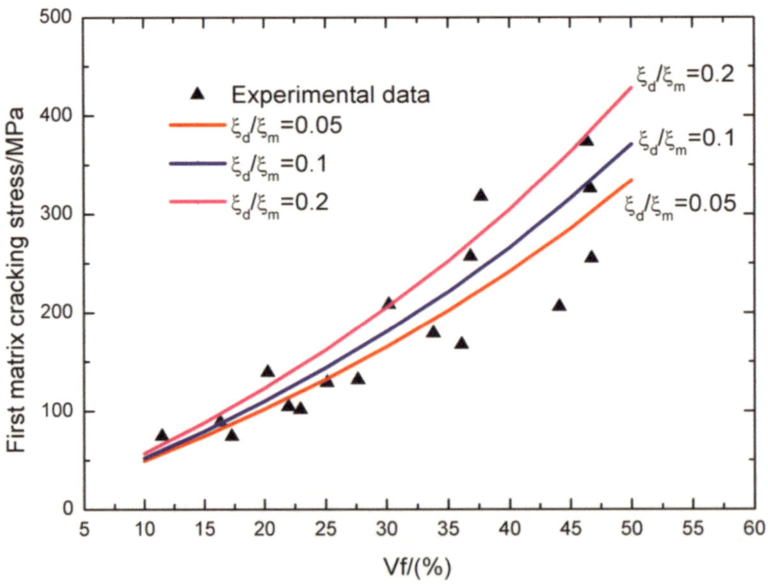

Figure 9.
The experimental and predicted matrix cracking stress versus the fiber volume corresponding to different fiber/matrix interface debonding energy of SiC/borosilicate composite.

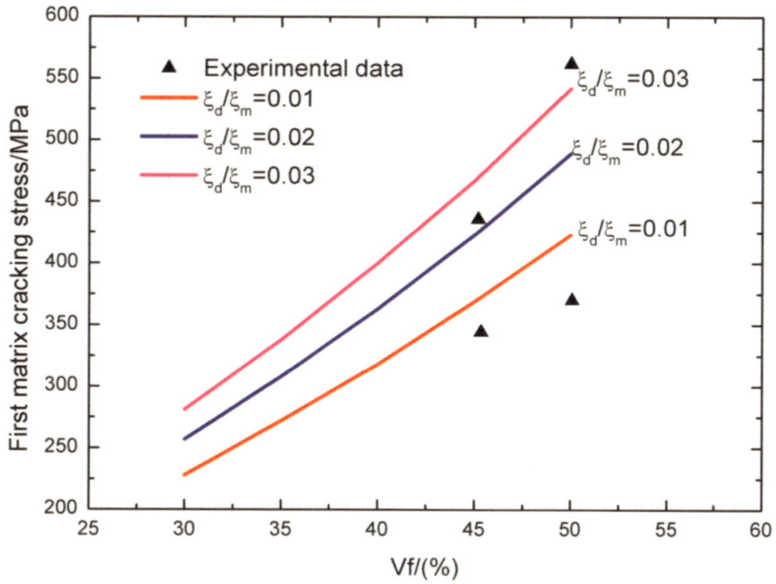

Figure 10.
The experimental and predicted matrix cracking stress versus the fiber volume corresponding to different fiber/matrix interface debonding energy of SiC/LAS composite.

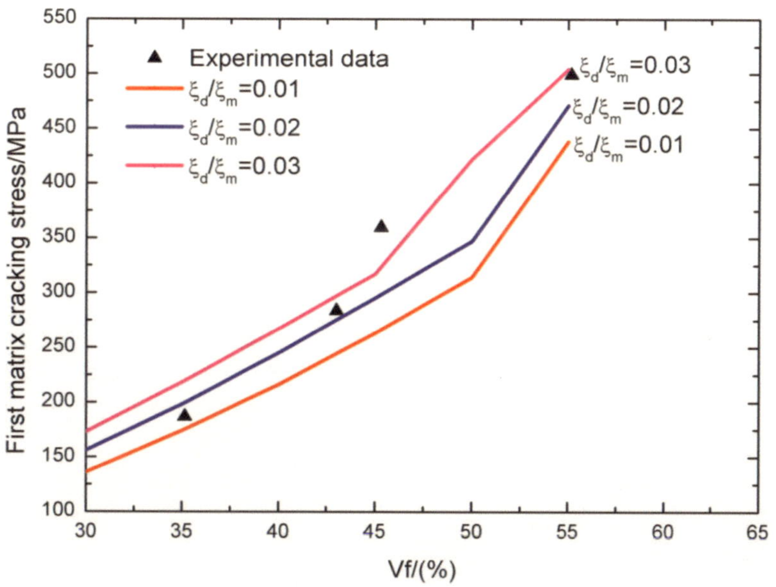

Figure 11.
The experimental and predicted matrix cracking stress versus the fiber volume corresponding to different fiber/matrix interface debonding energy of C/borosilicate composite.

agrees with the experimental data corresponding to the fiber volume changing from 30 to 55%, as shown in **Figure 11**. When the fiber/matrix interface debonding energy is ξ_d/ξ_m = 0.01, the matrix cracking stress increases from σ_{mc} = 136 MPa at the fiber volume of V_f = 30% to σ_{mc} = 438 MPa at the fiber volume of V_f = 55%; when the fiber/matrix interface debonding energy is ξ_d/ξ_m = 0.02, the matrix cracking stress increases from σ_{mc} = 156 MPa at the fiber volume of V_f = 30% to σ_{mc} = 471 MPa at the fiber volume of V_f = 55%; and when the fiber/matrix interface debonding energy is ξ_d/ξ_m = 0.03, the matrix cracking stress increases from σ_{mc} = 173 MPa at the fiber volume of V_f = 30% to σ_{mc} = 504 MPa at the fiber volume of V_f = 55%.

Yang [16] investigated the mechanical behavior of C/SiC composite after unstressed oxidation at elevated temperature of 700°C in air atmosphere. The composite was divided into two types based on the interface bonding, i.e., strong interface bonding and weak interface bonding. For the C/SiC with the strong interface bonding, the matrix cracking stresses of C/SiC corresponding to the proportional limit stresses in the tensile curves are 37, 30, and 20 MPa corresponding to the cases of without oxidation, 4 h oxidation, and 6 h unstressed oxidation, respectively. For the C/SiC with the weak interface bonding, the matrix cracking stresses

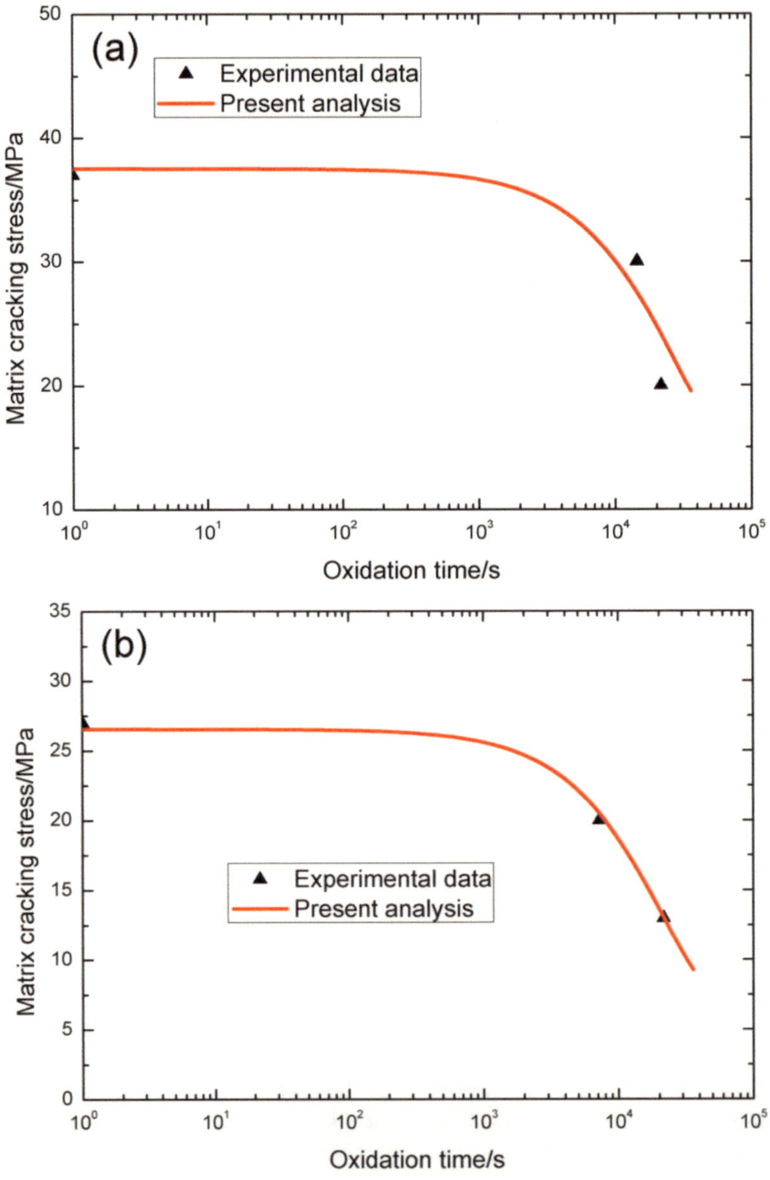

Figure 12.
The experimental and predicted matrix cracking stress versus the oxidation time of C/SiC composite after unstressed oxidation at 700°C in air corresponding to (a) strong interface bonding and (b) weak interface bonding.

of C/SiC corresponding to the proportional limit stresses in the tensile curves are 27, 20, and 13 MPa corresponding to the cases of without oxidation, 2 h oxidation, and 6 h unstressed oxidation, respectively. The experimental and predicted matrix cracking stresses of C/SiC composite with strong and weak interface bonding after unstressed oxidation at elevated temperature of 700°C in air are shown in **Figure 12a** and **b**, respectively. The matrix cracking stress decreases 18.9% after oxidation for 4 h, 46% after oxidation for 6 h for C/SiC with strong interface bonding, 25.9% after oxidation for 1 h, and 51.8% after oxidation for 6 h for C/SiC with weak interface bonding. The theoretical predicted results agreed with experimental data. The strong interface bonding can be used for oxidation resistant of C/SiC composite at elevated temperature.

5. Conclusion

In this chapter, the time-dependent matrix cracking of fiber-reinforced CMCs was investigated using the energy balance approach. The relationship between the matrix cracking stress, fiber and interface oxidation, and fiber failure was established. The effects of the fiber volume, interface shear stress and interface debonding energy, fiber failure, and oxidation temperature on the time-dependent matrix cracking stress were analyzed. The experimental matrix cracking stress of different fiber-reinforced CMCs was predicted.

1. When the fiber volume, fiber/matrix interface debonding energy and interface shear stress increase, the matrix cracking stress and the fiber/matrix interface oxidation length increase, and the fiber/matrix interface debonding length decreases.

2. When the fiber strength increases, the matrix cracking stress and the fiber/matrix interface debonding length increase, and the fiber/matrix interface oxidation length decreases.

3. When the oxidation temperature increases, the matrix cracking stress decreases, and the fiber/matrix interface oxidation length and the interface debonding length increase.

Acknowledgements

The work reported here is supported by the Fundamental Research Funds for the Central Universities (Grant No. NS2019038).

Author details

Li Longbiao
College of Civil Aviation, Nanjing University of Aeronautics and Astronautics, Nanjing, P.R. China

*Address all correspondence to: llb451@nuaa.edu.cn

References

[1] Li L. Damage, Fracture and Fatigue of Ceramic-Matrix Composites. Singapore: Springer Nature Singapore Pte Ltd.; 2018. DOI: 10.1007/978-981-13-1783-5. ISBN: 978-981-13-1782-8

[2] Li L. Thermomechanical Fatigue of Ceramic-Matrix Composites. New Jersey, USA: Wiley-VCH; 2019. ISBN: 978-3-527-34637-0. Available from: https://onlinelibrary.wiley.com/doi/book/10.1002/9783527822614

[3] Aveston J, Cooper GA, Kelly A. Single and multiple fracture. In: The Properties of Fiber Composites, Conference on Proceedings. National Physical Laboratory, Guildford: IPC Science and Technology Press; 1971. pp. 15-26

[4] Budiansky B, Hutchinson JW, Evans AG. Matrix fracture in fiber-reinforced ceramics. Journal of the Mechanics and Physics of Solids. 1986; **34**(2):167-189

[5] Rajan VP, Zok FW. Matrix cracking of fiber-reinforced ceramic composites in shear. Journal of the Mechanics and Physics of Solids. 2014;**73**:3-21

[6] Li L. Modeling first matrix cracking stress of fiber-reinforced ceramic-matrix composites considering fiber fracture. Theoretical and Applied Fracture Mechanics. 2017;**92**:24-32

[7] Li L. Time-dependent damage and fracture of fiber-reinforced ceramic-matrix composites at elevated temperatures. Composite Interfaces. 2019;**26**(11):963-988. DOI: 10.1080/09276440.2019.1569397

[8] Marshall DB, Cox BN, Evans AG. The mechanics of matrix cracking in brittle matrix composites. Acta Metallurgica et Materialia. 1985;**33**(11):2013-2021

[9] McCartney LN. Mechanics of matrix cracking in brittle-matrix fiber-reinforced composites. Proceedings of the Royal Society of London. 1987;**A409**:329-350

[10] Halbig MC, McGuffin-Cawley JM, Eckel AJ, Brewer DN. Oxidation kinetics and stress effects for the oxidation of continuous carbon fibers within a microcracked C/SiC ceramic matrix composite. Journal of the American Ceramic Society. 2008;**91**(2):519-526

[11] Casas L, Martinez-Esnaola JM. Modelling the effect of oxidation on the creep behavior of fiber-reinforced ceramic matrix composites. Acta Materialia. 2003;**51**(13):3745-3757

[12] Lara-Curzio E. Analysis of oxidation-assisted stress-rupture of continuous fiber-reinforced ceramic matrix composites at intermediate temperatures. Composites: Part A. 1999; **30**:549-554

[13] Curtin WA. Theory of mechanical properties of ceramic-matrix composites. Journal of the American Ceramic Society. 1991;**74**:2837-2845

[14] Gao YC, Mai YW, Cotterell B. Fracture of fiber-reinforced materials. Journal of Applied Mathematics and Physics. 1988;**39**(4):550-572

[15] Barsoum M, Kangutkar P, Wang A. Matrix cracking initiation in ceramic matrix composites, part I experiment and test results. Composites Science and Technology. 1992;**44**:257-269

[16] Yang CP. Mechanical characterization and oxidation damage modeling of ceramic matrix composites [PhD Thesis]. Xi'an: Northwestern Polytechnical University; 2011

Chapter 4

Effect of Ceramic/Graphite Reinforcement on Dry Sliding Wear Behavior of Copper Metal Matrix Hybrid Composites

Manvandra Kumar Singh, Mulkraj Anand, Pushkar Jha and Rakesh Kumar Gautam

Abstract

In the present investigation, effect of ceramic/graphite addition on the dry sliding wear behavior of copper-based hybrid composites have been assessed at three different normal loads of 9.81, 19.62 and 29.43 N. Wear test is performed by using pin-on-disc test rig at sliding speeds of 1.30 and 1.84 m/s under ambient conditions. The copper-based hybrid composites are successfully synthesized by using stir casting route. The samples are characterized by various techniques such as X-rays diffraction (XRD), high resolution-scanning electron microscopy (HR-SEM), scanning electron microscope (SEM) equipped with energy dispersive analysis of X-ray spectroscopy (EDAX). Microstructural investigations reveal the presence of the WC, TiC and graphite particles in the hybrid composites. It is observed that graphite reinforced hybrid composite shows better wear resistance than hybrid composite without graphite addition. In all the cases cast copper specimen shows highest wear rate. The observed friction and wear behavior have been explained on the basis of the presence tribofilm on the worn surface. Microstructural characterization of the worn surface and wear debris show that the mechanism of wear is primarily adhesive and oxidative in case of cast copper, where as it is a mix of adhesive and abrasive wear in case of hybrid composites. Wear debris analysis also helps to understand the wear mechanism involved during dry sliding.

Keywords: metal matrix composites, friction, wear, tribofilm, dry-sliding

1. Introduction

During the last two decades, metal matrix composites (MMCs) have gained researcher's focus due to its outstanding properties such as high elastic modulus, specific strength, specific stiffness, desirable coefficient of thermal expansion, high temperature resistance, and better wear resistance [1, 2]. Among the existing metal matrix composites, copper based composites are gaining enormous importance due to its various automobile and industrial applications viz. automobile radiators, bearings and bushings, heat exchangers, electrical sliding contacts and resistance welding electrodes [3–5]. Copper based composites can be developed by using various techniques viz. stir casting, powder metallurgy, chemical vapor deposition, etc. However, among these techniques, stir casting technique has an edge over other fabrication techniques [6]. Researchers have reinforced various ceramics such as SiC, TiC, WC, Al_2O_3, graphite and TiB_2, etc. into the copper matrix to enhance its mechanical and tribological properties [7–12]. Kumar and Mondal [13] prepared graphite reinforced copper matrix composites and explored its friction and wear properties at normal loads of 10–55 N and constant velocity of 0.77 m/s by using pin-on-disk. It was noticed that graphite reinforced composites showed improved wear resistance and lower friction coefficients compared to unreinforced copper. The friction and wear resistance improvements are mainly caused due to the formation of a solid self-lubricating film formed by the presence graphite particles. Gautam et al. [14] prepared Cu-Cr-graphite composite through solidification processing. Dry sliding test was conducted at different normal loads 10, 20, 30, and 40 N and a constant sliding speed of 0.786 m/s. It was observed that wear rate of graphite reinforced composites are lower than other composites. It was also observed that average coefficient of friction is reciprocal with the normal load and the graphite added composite reveals the lesser average coefficient of friction. Nayak et al. [15] developed graphite and TiC reinforced copper hybrid composites. It was noticed that hardness of the composites decreases with graphite addition while its hardness increases with TiC addition. Wear test was conducted using a pin- on-disk test rig. Worn surface investigation shows that the reinforcement of TiC and graphite considerably enhances the tribological properties of the copper composites. Rajkumar and Aravindan [16] revealed the tribological behavior of Cu-TiC-graphite hybrid composites under the conditions (normal loads of 12–48 N; and sliding speeds of 1.25–2.51 m/s). It was noticed that the addition of higher graphite content helps to form a mixed smooth layer that enhances the wear resistance and decreases the friction coefficient.

From the literature review it can be concluded that several researchers have studied and reported the various mechanical and tribological behavior of Cu-based hybrid composites. However, only few reports are available for exploring the friction and wear behavior of the graphite along with two ceramics (WC and TiC) reinforced in the copper matrix. Hence, in the present investigation the main focus is to explore the dry sliding wear behavior of graphite reinforced Cu-based hybrid composites by using pin on disk test rig. Further, various characterization techniques were employed to examine the worn surface and wear debris morphologies.

2. Experimental details

Small pieces of commercially (99.99%) pure copper in calculated amount were taken in a graphite crucible and placed in the electric muffle furnace. The temperature of the furnace was set to 1200°C with heating rate of 300°C/h and holding time at this temperature of furnace was set for 30 min. As the furnace temperature attained the temperature of 1200°C, copper pieces in graphite crucible completely changed into molten metal state and then fixed amount of chromium (2 wt%) was added into the melt. Reinforcements such as WC, TiC and graphite in calculated amounts as given in **Table 1** were poured into the molten metal by wrapping in copper foil as small packets with the help of tong and continuously stirred at 300 rpm for 5 min. **Figure 1** presents the SEM morphology of the reinforcing particles (WC, TiC and graphite). **Figure 1(a)** and **(b)** reveals the equiaxed, regular shape and size of WC and TiC. **Figure 1(c)** presents the equiaxed and flake type geometry of the graphite particles. Reinforcements were preheated at temperature of 100°C to make it moisture free. Three compositions were obtained after casting; (a) cast copper, (b) Cu-2 wt% Cr-1.5 wt% WC-1.5 wt% TiC and (c) Cu-2 wt% Cr-1.5 wt%

S. no.	Materials	Compositions	Wt%	Minimum assay	Wt in (g)	Wt in (g) after assay correction
1.	Cast commercial copper	Commercial copper (CC)	100.0	99.5%	1000.0	1005.0
2.	Hybrid composite-1 (HC-1)	Commercial copper (CC)	95.0	99.5%	950.0	954.7
		Tungsten carbide (WC)	1.5	98.0%	15.0	15.3
		Titanium carbide (TiC)	1.5	98.5%	15.0	15.2
		Chromium (Cr)	2.0	99.0%	20.0	20.2
3.	Hybrid composite-2 (HC-2)	Commercial copper (CC)	90.0	99.5	900.0	904.5
		Titanium carbide (TiC)	1.5	98.0	15.0	15.3
		Tungsten carbide (WC)	1.5	98.5	15.0	15.2
		Chromium (Cr)	2.0	99.0	20.0	20.2
		Graphite	5.0	98.0	50.0	51.0

Table 1.
Compositional details of the developed materials.

Figure 1.
SEM micrographs of as received (a) WC, (b) TiC and (c) graphite powders.

WC-1.5 wt% TiC-5 wt% Gr and these were designated as CC, HC-1, and HC-2, respectively. After stirring of the molten materials, it was poured into a permanent steel mold and kept for air cooling. After cooling, the developed cast ingot of matrix and hybrid composites were used for different characterizations.

The phase evolution of the hybrid composites were studied by using Rigaku miniflex600 with DTEX-ultra diffractometer with Cu-Kα radiation (λ = 1.54A°). High resolution scanning electron microscope (FEI Netherlands, NOVA NANOSEM450) and scanning electron microscope (ZEISS-EVO/18) equipped with EDAX were utilized for microstructural examinations. Brinell hardness test was performed as per ASTM E10-00 standards, using a load of 500 kg and a 10 mm ball indenter at room temperature. The average of five measurements was reported to ascertain reproducibility.

Friction and wear test of both cast copper and hybrid composites were conducted according to the ASTM G99-05 standards using standard pin on disc wear test rig (MAGNUM Bangalore, India) with a counterface of the EN31 steel hardened to 60 ± 3 HRC at an ambient condition. The samples on the test rig, the surface of the hardened steel disc and test pin were thoroughly cleaned with emery paper and acetone to remove any possible traces of grease and other surface contaminants.

The normal loads were 9.81, 19.62, and 29.43 N and the sliding speeds were 1.30 and 1.84 m/s, respectively. The total sliding distance covered during the wear test was ~5000 and ~7000 m. The weight loss was measured prior to and after the test using an electronic balance with an accuracy of 0.0001 mg. Meanwhile, the volume loss of the samples was calculated through weight loss divided by Archimedes' density of the material. The control panel equipped with the tribometer displayed the frictional force. The same was used to calculate the friction coefficient by dividing it by the normal load. Each test for a particular condition of load and speed was conducted thrice. The average value of three tests was reported in the present investigation. The worn surface of the specimens and wear debris were subjected to SEM and EDAX analyses to explore the nature of the wear mechanism.

3. Results and discussion

3.1 X-ray diffraction (XRD)

X-ray diffraction (XRD) pattern of the CC, HC-1 and HC-2 is shown in **Figure 2**. The diffraction peaks present in the developed materials are indexed with XRD-JCPDS database. One can observe the presence of CuO and Cu_2O phases in the XRD spectra and it can be possibly due to the oxidation of the copper at high temperature with residual atmospheric oxygen present in the furnace. However, the peaks of WC and TiC are not distinct and it can be attributed to the lower weight percentage of the reinforcing particles in the copper matrix which may not be in the detactable range of XRD [17].

3.2 Microstructure

Figure 3(a)–(c) shows the HR-SEM microstructure of the CC, HC-1 and HC-2, respectively. **Figure 3(a)** depicts the microstructure of CC, in which the presence of any foreign particles could not be observed. It shows a plane, smooth and homogeneous morphology. However, the microstructure of hybrid composites show some embedded reinforcing particles into the metal matrix. The embedded reinforcing particles resemble with the crystal geometry of WC which is hexagonal and cubic structure of TiC particles, respectively [18]. The typical microstructures of hexagonal and cubic crystal geometry can be observed in **Figure 3(b)**. **Figure 3(c)** shows the microstructure of the HC-2 which depicts the presence of reinforcing particles into the metal matrix that is shown in the inset. The inset particle are magnified and analyzed for its microstructural geometry. It is found that hexagonal and cubical crystal geometry are particularly the crystal geometry of WC and TiC in the matrix as shown in **Figure 3(c-1)** and **(c-2)**, respectively.

Figure 4 presents the EDAX spectrum of developed materials. The EDAX spectrum of the CC, HC-1 and HC-2 are shown in the **Figure 4(a-a′)**, **(b-b′)** and **(c-c′)**, respectively. Very weak intensity peak of carbon and oxygen are observed in the EDAX spectrum of CC as shown in **Figure 4(a′)**. Carbon intensity peak may be attributed to the presence of less impurity of carbon in CC and oxygen intensity peak may be due to the possible oxidation of the CC at high temperature during melting. Atomic and weight percentage of carbon, oxygen and copper are shown in the corresponding table in **Figure 4**. However, **Figure 4(b′)** and **(c′)** show the EDAX spectrum of the HC-1 and HC-2, respectively. From spectrum, comparatively smaller intensity peak of carbon, oxygen, tungsten, titanium and chromium are observed than that of the copper intensity peak. It can be observed that atomic/ weight percentage of carbon and oxygen found in hybrid composites are higher

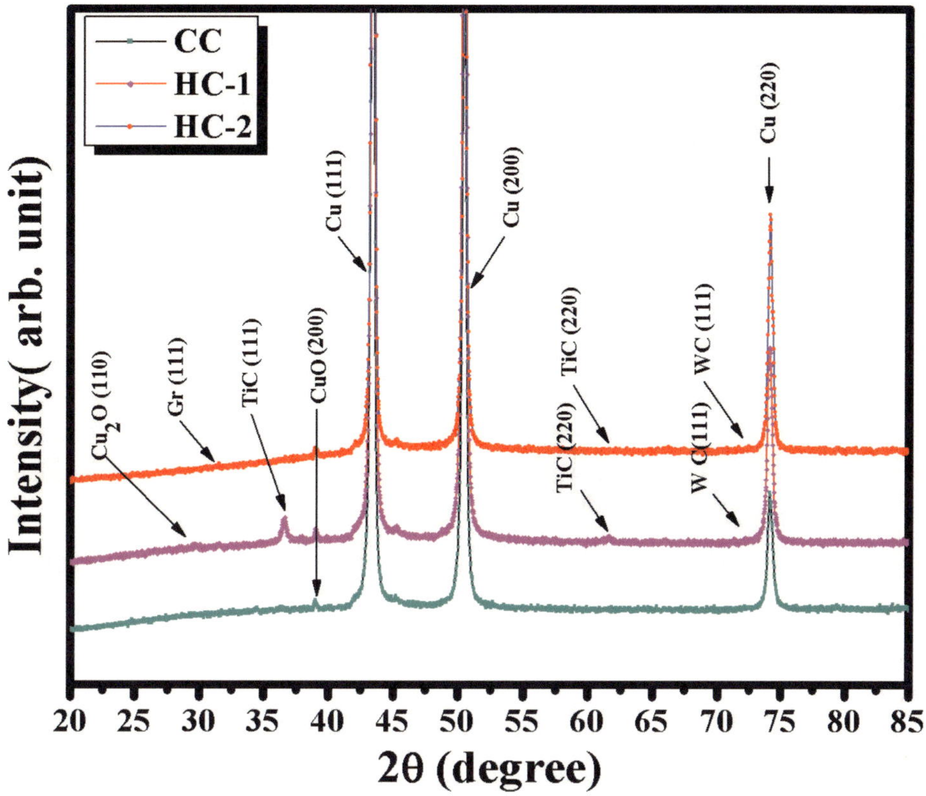

Figure 2.
X-ray diffraction patterns of CC, HC-1 and HC-2 specimens.

than the CC as depicted in the table of **Figure 4(b′)** and **(c′)**. The possible reason for the increased atomic/weight percentage of carbon in EDAX spectrum of composites can be due to the addition of carbon in form of carbides and graphite. It can be concluded from the **Figure 4(b′)** and **(c′)** that the WC, TiC and graphite are added to the commercial copper.

3.3 Density and hardness

The variation in the density of the developed materials is shown in **Figure 5**. It is clear from the figure that HC-1 shows higher density as compared to the CC and HC-2. The reason for the higher density of HC-1 can be credited to the improved bonding between the reinforcement particles and the copper matrix. To enhance the bonding in the hybrid composites, Cr has been used as a wetting agent [14]. It can also be inferred that HC-2 shows lower density than cast copper and HC-1. The possible reason for such lowering in density of HC-2 can be due to the reinforcement of ultralow density graphite particles [19].

Figure 5 also depicts the variation in the hardness of CC, HC-1 and HC-2. It is obvious from the figure that hybrid composites (HC-1 and HC-2) show higher hardness as compared to the cast copper matrix. The reason for the increased hardness of HC-1 and HC-2 can be attributed to the reinforcement of harder WC and TiC ceramics in the ductile copper matrix. During hardness test, when the load is applied through ball indentation on the materials, the applied load transfers from the softer matrix to the harder reinforcing particle and it restricts further plastic deformation. The differences in the elastic modulus and coefficient of thermal expansion of matrix and reinforcement can also have a significant

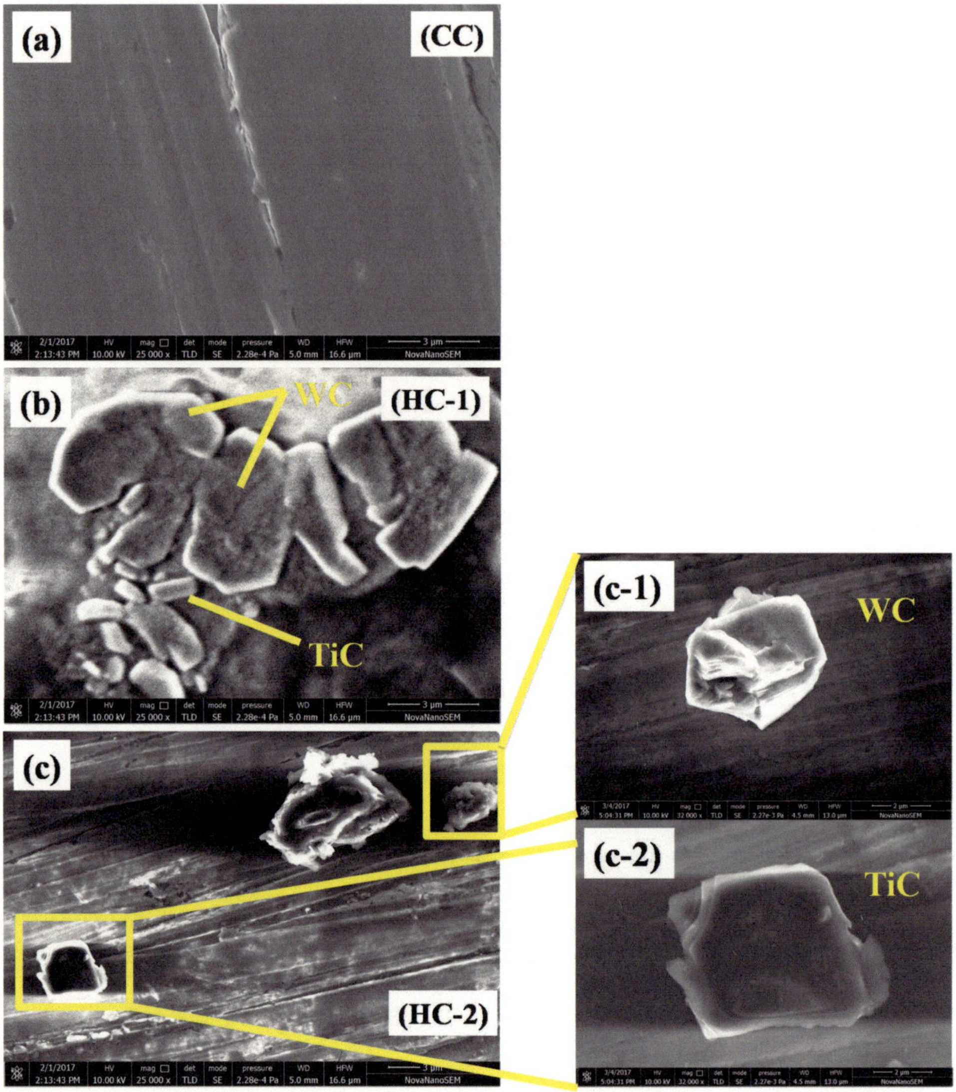

Figure 3.
HR-SEM microstructure of developed materials (a) CC, (b) HC-1, and (c) HC-2.

role in the improvement of hardness of the hybrid composites. In addition, the strain hardening also affects the hardness behavior of the materials [20]. It is observed that HC-1 and HC-2 shows almost 80 and 50% increment in hardness as compared to the copper matrix, respectively. The hardness of the HC-2 is lower as compared to HC-1; this is due to the reinforcement of softer graphite materials in the softer matrix [19].

3.4 Investigation of sliding wear behavior

Figure 6(a)–(f) shows the variation of the cumulative volume loss with the sliding distance for all the materials investigated in the present study under different loads of 9.81, 19.62 and 29.43 N for sliding speed of 1.30 and 1.84 m/s. It can be observed that cumulative volume loss increases as the sliding distance increases for CC, HC-1 and HC-2, respectively. Moreover, the volume loss of CC is higher at all the normal loads as compared to the hybrid composites. Lower volume loss of the

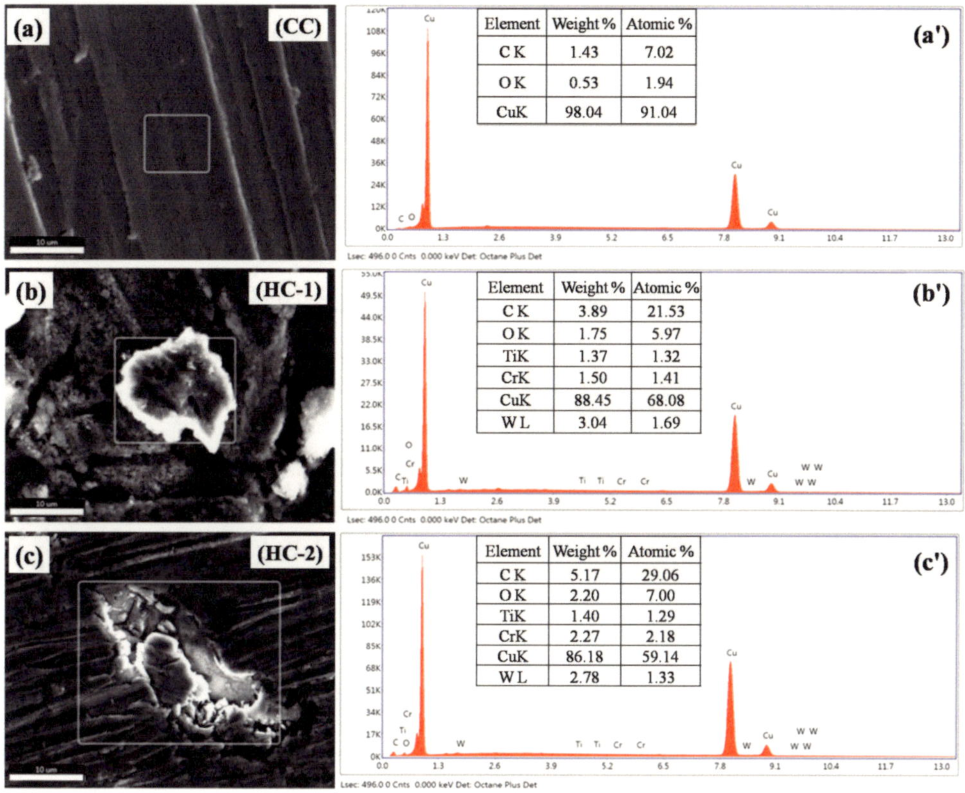

Figure 4.
EDAX spectrum of the marked portion of the micrograph of (a) CC, (b) HC-1, and (c) HC-2.

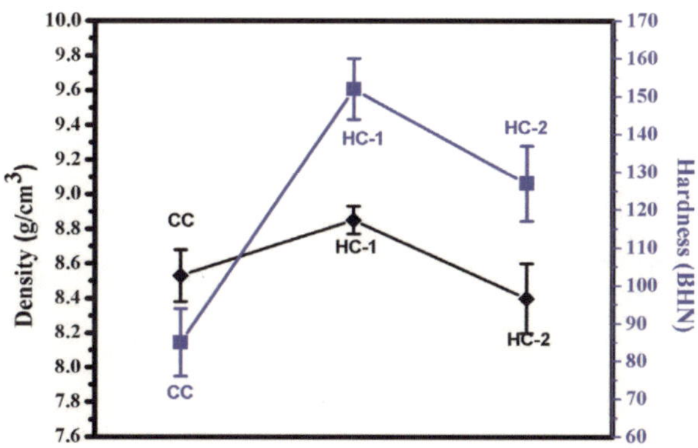

Figure 5.
Variation of density and hardness of the developed materials.

hybrid composites can be credited to the reinforcement of harder particles (TiC and WC) in case of HC-1 because these harder particles provide a shield to the relatively softer matrix during sliding and enhance the load bearing capacity of the hybrid composite [21]. In case of HC-2, the volume loss is minimum and it can be attributed to the combined effect of harder phase (TiC and WC) as well as softer phase (graphite) reinforcement. During sliding, the graphite particles comes out from the surface of the wear pin and act as solid lubricant and develop a thin tribofilm between the wear pin and hard counter surface which reduces the coefficient of friction thus it reduces the volume loss.

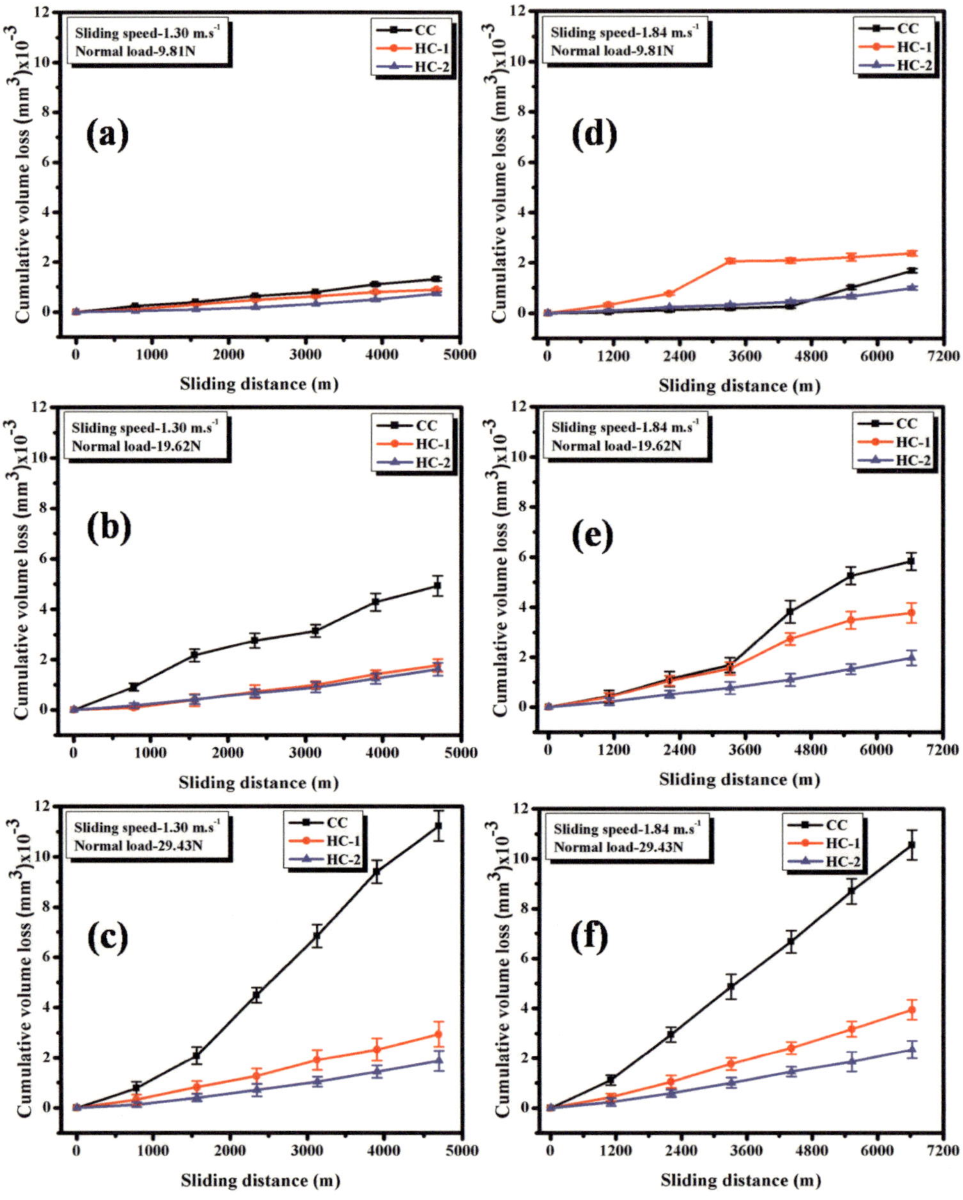

Figure 6.
Variation of cumulative volume loss with sliding distance for CC, HC-1 and HC-2 at 1.30 m/s and at 1.84 m/s.

Figure 7(a) and **(b)** shows the variation of the wear rate with normal load for CC, HC-1 and HC-2 at sliding speed of 1.30 and 1.84 m/s. It is clear from **Figure 7(a)** and **(b)** that the wear rate increases as the load increases at both the sliding speeds. Among the materials investigated, CC shows the highest wear rate and HC-2 shows lowest wear rate. In case of CC, there are chances of direct metal to metal contact and it may lead to adhesive wear. Adhesive wear mechanism initiates plastic deformation which is responsible for the highest wear rate. Similar observations for wear behavior of unreinforced copper were reported by Tu et al. [22]. The wear rate of HC-1 is comparatively lesser than CC because of reinforcement of harder particle like—WC and TiC. During sliding contacts these harder particles get early exposure on the surface of the wear pin and these particles may carry the applied normal load for a longer time without transferring the contacts to copper matrix [23].

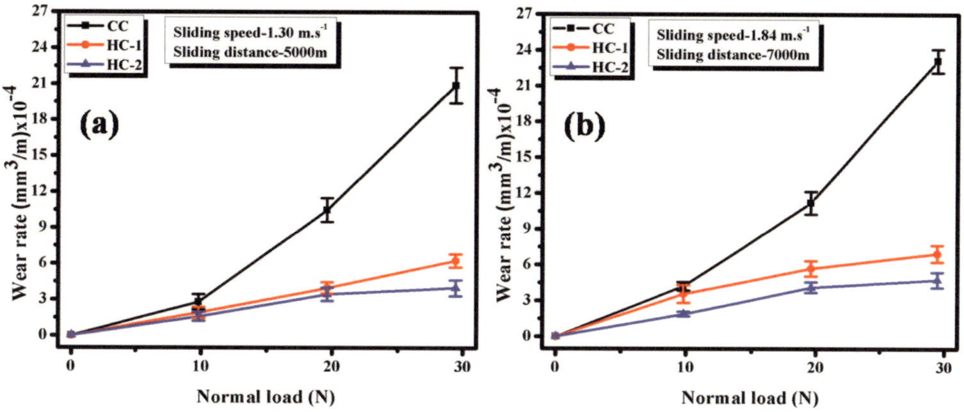

Figure 7.
Variation of wear arte with normal load for CC, HC-1 and HC-2 at (a) 1.30 m/s and (b) at 1.84 m/s.

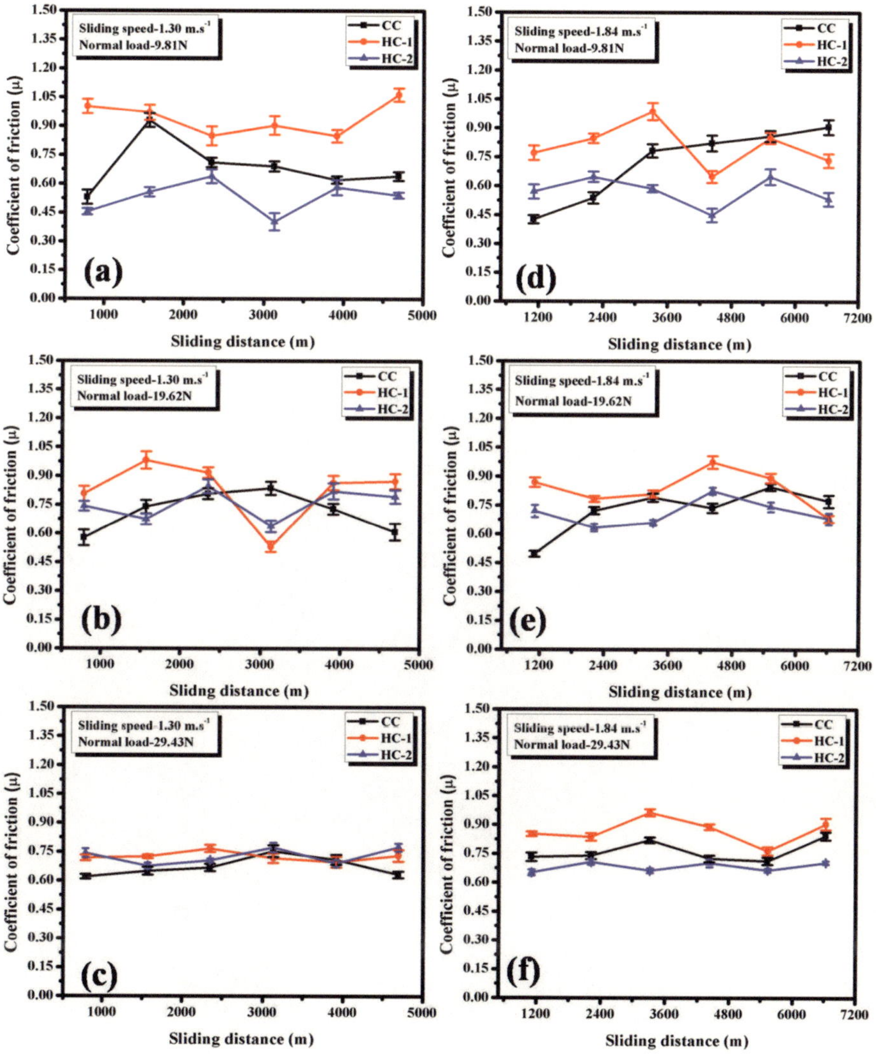

Figure 8.
The variation of coefficient of friction with sliding distance of CC, HC-1 and HC-2 at 1.30 m/s (a)–(c) and at 1.84 m/s (d)–(f).

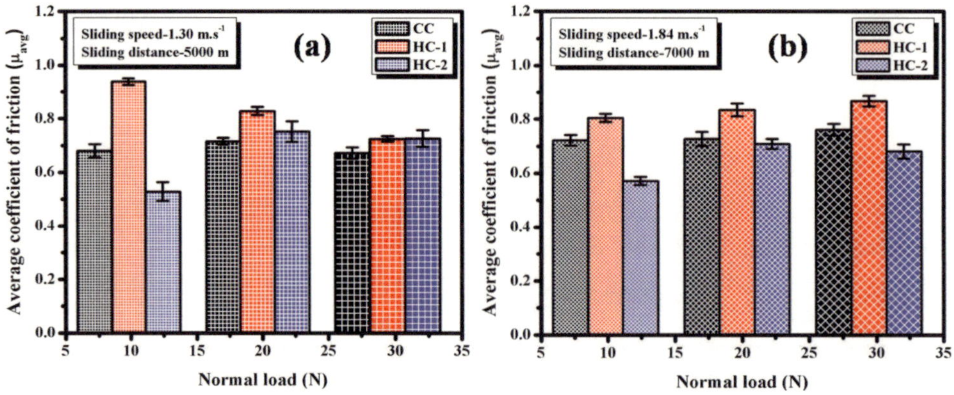

Figure 9.
The variation of the average coefficient of friction with normal load for CC, HC-1 and HC-2 at (a) 1.30 m/s and (b) 1.84 m/s.

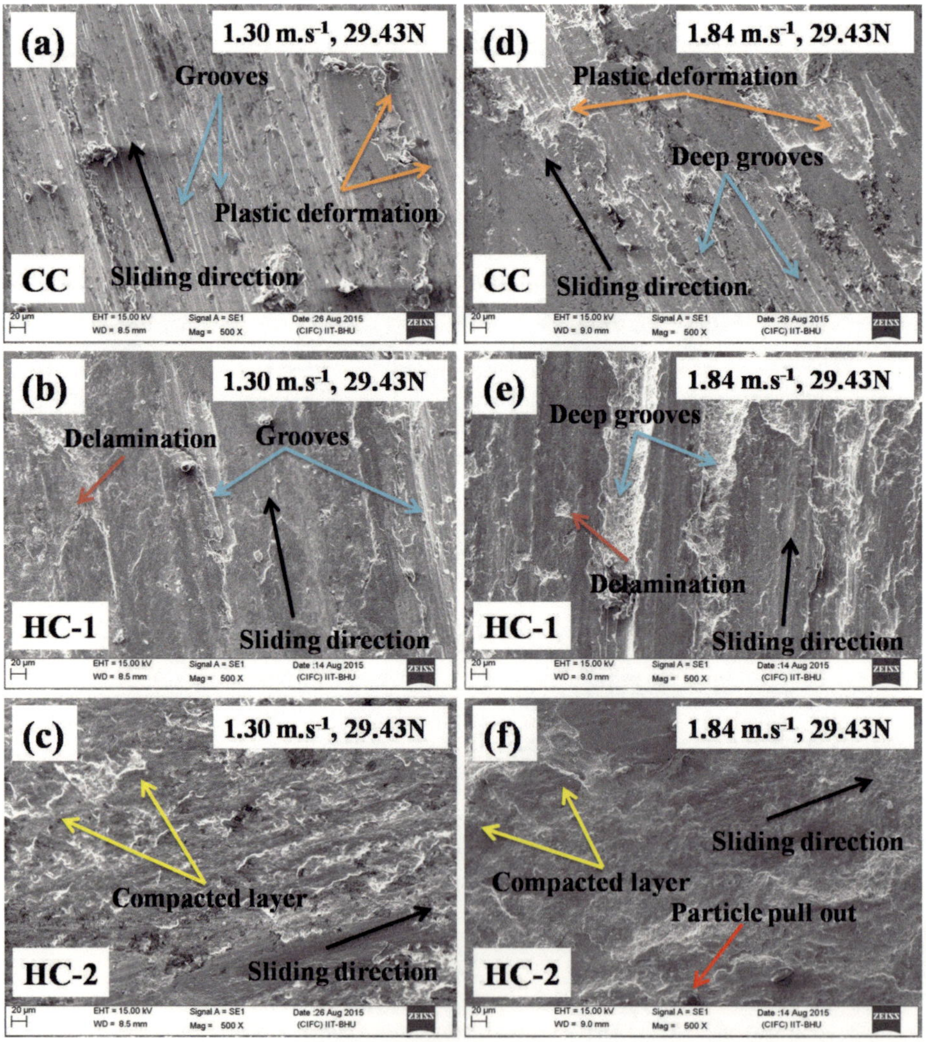

Figure 10.
SEM micrograph of worn surfaces at 29.43 N of (a) CC, (b) HC-1, (c) HC-2 at 1.30 m/s and (d) CC, (e) HC-1, (f) HC-2 at 1.84 m/s.

The wear mechanism changes from adhesive to abrasive nature. It can be observed that HC-2 shows lowest wear rate and the reason can be credited to the combined effect of hard particles reinforcement and squeezing out of the graphite particle. At the highest load the debris particles that comes during sliding, gets compacted between the mating surfaces. This compacted layer of debris which contains the graphite particles help to form a type of tribofilm and it may be responsible for the lowest wear rate [16]. It can be observed from the **Figure 7** that as sliding speed increases from 1.30 to 1.84 m/s the wear rate increases for CC, HC-1 and HC-2. The reason can be attributed to the fact that as sliding speed increases it gives rise to increase in frictional heat energy. The dissipated heat energy makes the matrix surface softens and weakens the interface with allowing easier pull out of the particulates [24].

The typical variation of coefficient of friction with sliding distance for CC, HC-1 and HC-2 at different normal loads (9.81, 19.62 and 29.43 N) and sliding speeds (1.30 and 1.84 m/s) are presented in **Figure 8(a)–(f)**. It is observed that coefficient of friction fluctuates with the sliding distance. The reason for this fluctuation can be due to the disparity in contact that occurs when the sample and the counterface are evolving to develop a better surface conformity. Other reason for the fluctuation of hybrid composites can be credited to the fact that during sliding the abrasive particles comes between the contacting surfaces.

Figure 9(a) and **(b)** show the variation of the average coefficient of friction with normal load for CC, HC-1 and HC-2 at sliding speeds of 1.30 and 1.84 m/s,

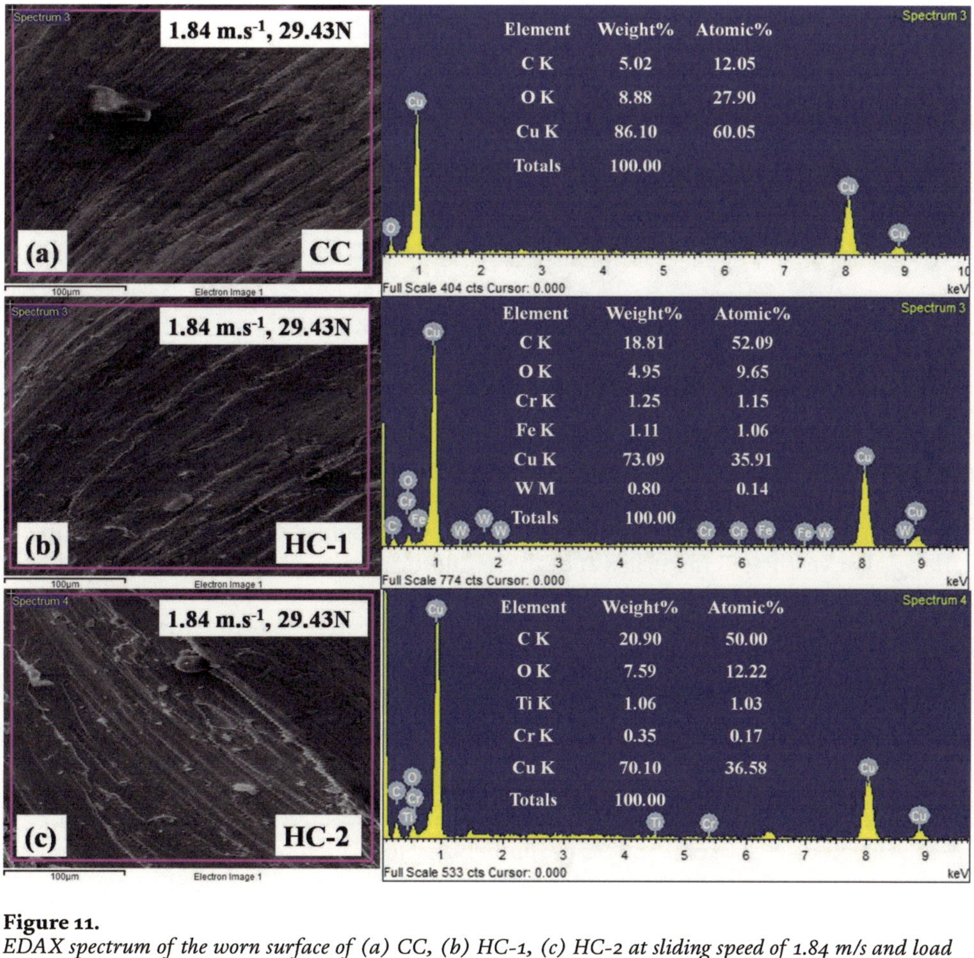

Figure 11.
EDAX spectrum of the worn surface of (a) CC, (b) HC-1, (c) HC-2 at sliding speed of 1.84 m/s and load of 29.43 N.

Figure 12.
SEM morphology of wear debris at 29.43 N of (a) CC, (b) HC-1, (c) HC-2 at 1.30 m/s and (d) CC, (e) HC-1, (f) HC-2 at 1.84 m/s.

respectively. It can be observed that HC-1 shows the highest coefficient of friction among all investigated materials at both the sliding speeds. It can also be observed that coefficient of friction increases as load increases from 9.81 to 29.43 N. The reason for the highest coefficient of friction of HC-1 can be credited to the reinforcement of ceramic particles and these particles when comes between the contacting surfaces may give rise to three body abrasion and thus coefficient of friction increases [25]. As the load increase the surface contact increases and thus more surface asperities come in contact and it can be the reason for the increase in the coefficient of friction for the materials. However, one can observe that HC-2 shows lower coefficient of friction than CC and HC-1. The possible reason for this can be due to the presence of softer phase graphite particles which reduces the coefficient of friction in all the cases due to the formation of the tribofilm. The tribofilm works as a lubricating layer and reduces the coefficient of friction with the sliding distance [26].

Figure 10(a)–(f) show the SEM micrographs of worn surfaces of CC, HC-1 and HC-2 at maximum normal load of 29.43 N and sliding speeds of 1.30 and 1.84 m/s,

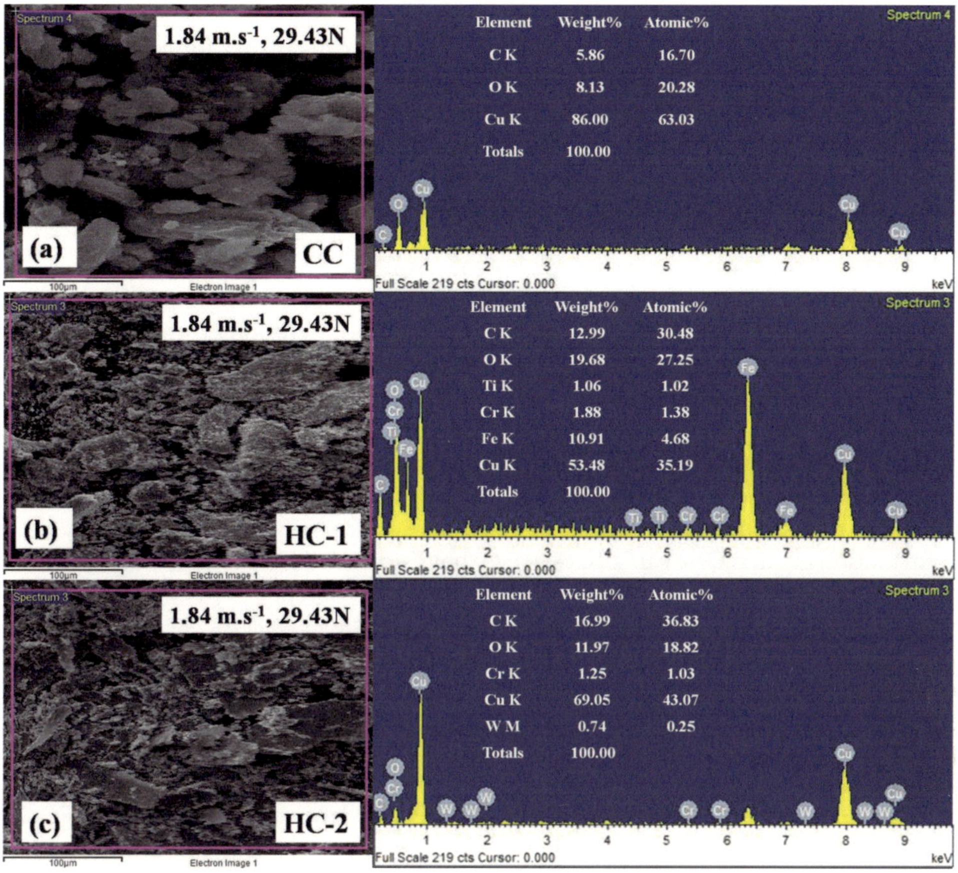

Figure 13.
EDAX spectrum of wear debris of (a) CC, (b) HC-1 and (c) HC-2 at sliding speed of 1.84 m/s and load of 29.43 N.

respectively. The corresponding SEM micrographs of the CC worn surface shown in **Figure 10(a)** and **(d)** exhibit the presence of the grooves due to the adhesive wear mechanism which attributed to the direct metal to metal contact. Due to the adhesive wear plastic deformation can be observed at the worn surfaces. One can observe that grooves are less deep in **Figure 10(a)** at lower sliding speed of 1.30 m/s as compared the grooves appearing in **Figure 10(d)** at higher sliding speed of 1.84 m/s. **Figure 10(b)** and **(e)** show the worn surface of HC-1 at both the sliding speeds. The grooves along with some delamination can be observed in the micrograph shown in **Figure 10(b)**. Similarly, **Figure 10(e)** shows the deeper grooves with delamination at higher sliding speed of 1.84 m/s. The micrographs predict that there is transformation of wear mechanism from adhesive to abrasive wear in case of HC-1 hybrid composite. The abrasive actions are mainly caused by the wear debris that mainly contains traces of reinforced ceramic particles. Due to the abrasive wear mechanism in case of HC-1 as shown in the **Figure 10(b)** and **(e)** it can be confirmed that the results given in **Figure 9** are convincing. The worn surface of HC-2 is shown in **Figure 10(c)** and **(f)** which shows less grooves and mild wear region as compared to CC and HC-1at both the sliding speeds. The presence of lesser or no grooves on the worn surfaces at both the sliding speeds for HC-2 can be due to the reason that these worn surfaces may be covered by the transfer layer as depicted in the micrograph. This layer is basically formed from the wear debris coming out during sliding. When this transfer layer gets compacted during sliding, because heat is generated during sliding, then it protects the surface against wear and thus wear rate decreases as shown in **Figure 7** [27]. In **Figure 10(f)**, particles pullout can be seen which comes from the wear pin during sliding and these particles may create a lubricating surface which comes between the sliding surfaces and defends the surface against wear. This layer also inhibits a direct metal–metal contact and provides a low shearing strength junction at the interface, thereby resulting in a reduced friction coefficient in comparison to the hybrid composites as shown in **Figure 9**.

Figure 11(a)–(c) shows the typical EDAX spectrum of the marked portion in worn surface of CC, HC-1, HC-2 at sliding speed of 1.84 m/s and normal load of 29.43 N respectively. **Figure 11(b)** displays the presence of Fe in the EDAX spectra of HC-1which confirms that the metal transfer occurred from the counter face. The presence of oxygen in the spectra indicates the possibility of oxidation that might have occurred during sliding possibly due to increase in contacting surfaces.

The wear debris analysis helps to predict the wear mechanism involved during dry sliding. **Figure 12** shows SEM morphology of wear debris of (a) CC, (b) HC-1, (c) HC-2 at normal load of 29.43 N and at sliding speed of 1.84 m/s. It can be observed that wear debris particle appearing in **Figure 12(a)** are larger in size which got detached from the CC due to adhesive wear mechanism and thus plastic deformation occurs. One can also observe that **Figure 12(b)** which corresponds to the wear debris of HC-1 appears like plate and flake shaped. The reason can be attributed to an abrasive micro-cutting effect. The other cause for the smaller wear debris is due to the higher hardness of HC-1. **Figure 12(c)** shows equiaxed sheets and powder like morphology. Similar observations were reported by Sundararajan and Rajadurai [28]. This type of morphology predicts that wear is lesser and thus the debris particles are smaller. In this case it can be predicted that these fine debris particles acts as lubricant during sliding.

Figure 13 shows the EDAX spectrum of the wear debris of CC, HC-1 and HC-2 at normal load of 29.43 N and sliding speed of 1.84 m/s. EDAX spectrum of wear debris shows the intensity peak of entire reinforcing elements with their atomic and weight % that are shown in the spectrum. The greater amount of weight and atomic % of oxygen are observed in the entire EDAX spectrum of wear debris compared to EDAX spectrum of developed materials. Such increment in oxygen % in EDAX spectrum of wear debris indicates the involvement of oxidative wear of the materials during dry sliding. The intensity peak of iron element is observed in the EDAX spectrum of wear debris of HC-1 at both the sliding speeds. It is attributed to the higher hardness of the HC-1 which abrades the counter steel disc in lesser amount and adheres with the other wear debris. The analysis of EDAX spectrum of the wear debris confirms that mainly oxidative, abrasive and adhesive wear mechanism takes place during dry sliding.

4. Conclusions

- Copper-based hybrid composites have been successfully synthesized by using stir casting route. XRD pattern confirms that TiC, WC, and graphite are present in the hybrid composites.

- Microstructural analysis reveals the presence of reinforcing particles and its uniform distribution. Density of the HC-2 is lowest. However, the hardness of the HC-1 is found to be maximum.

- The volume loss of CC is higher at all the normal loads as compared to the hybrid composites. CC shows highest wear rate among the developed specimens. HC-2 shows better wear resistance than HC-1. This has been attributed to the lubricating effect of graphite that forms tribofilm between the contacting surfaces during sliding.

- It is observed that there are no certain trends of coefficient of friction with sliding distance for CC, HC-1 and HC-2 hybrid composites. It can also be observed that HC-2 shows lower coefficient of friction than CC and HC-1. The possible reason for this can be due to the presence of softer phase graphite particles which reduces the coefficient of friction in all the cases.

- Worn surface analysis predicts that the mechanism of wear is primarily adhesive and oxidative in case of CC where as it is a mix of adhesive and abrasive wear in case of hybrid composites.

- EDAX spectra confirm that the metal transfer occurred from the counter face in case of hybrid composites. The presence of oxygen in the spectra indicates the possibility of oxidation that might have occurred during sliding. Wear debris analysis also helps to understand the wear mechanism.

Author details

Manvandra Kumar Singh[1*], Mulkraj Anand[2], Pushkar Jha[3]
and Rakesh Kumar Gautam[2]

1 Department of Mechanical Engineering, Mewar University, Chittorgarh, India

2 Department of Mechanical Engineering, Indian Institute of Technology (BHU), Varanasi, India

3 School of Mechanical Engineering, KIIT Deemed to be University, Bhubaneswar, India

*Address all correspondence to: mksingh.rs.mec13@itbhu.ac.in

References

[1] Miracle DB. Metal matrix composites from science to technological significance. Composites Science and Technology. 2005;**65**:2526-2540

[2] Ralph B, Yuen HC, Lee WB. The processing of metal matrix composites—An overview. Journal of Materials Processing Technology. 1997;**63**:339-353

[3] Heringhaus F, Raabe D. Recent advances in the manufacturing of copper-base composites. Journal of Materials Processing Technology. 1996;**59**:367-372

[4] Sapate SG, Uttarwar A, Rathod RC, Paretkar RK. Analyzing dry sliding wear behaviour of copper matrix composites reinforced with pre-coated SiCp particles. Materials and Design. 2009;**30**:376-386

[5] Mai YJ, Chen FX, Lian WQ, Zhang LY, Liu CS, Jie XH. Preparation and tribological behavior of copper matrix composites reinforced with nickel nanoparticles anchored graphene nanosheets. Journal of Alloys and Compounds. 2018;**756**:1-7

[6] Hashim J, Looney L, Hashmi MSJ. Metal matrix composites: Production by the stir casting method. Journal of Materials Processing Technology. 1999;**92-93**:1-7

[7] Celebi Efe G, Ipek M, Zeytin S, Bindal C. An investigation of the effect of SiC particle size on Cu–SiC composites. Composites: Part B. 2012;**43**:1813-1822

[8] Akhtar F, Askari SJ, Shah KA, Du X, Guo S. Microstructure, mechanical properties, electrical conductivity and wear behavior of high volume TiC reinforced Cu-matrix composites. Materials Characterization. 2009;**60**:327-336

[9] Hong E, Kaplin B, You T, Suh M-s, Kim Y-S, Choe H. Tribological properties of copper alloy-based composites reinforced with tungsten carbide particles. Wear. 2011;**270**:591-597

[10] Besterci M, Kohú I, t, Velgosova O. Microstructural parameters of dispersion strengthened Cu–Al_2O_3 materials. Journal of Materials Science. 2008;**43**:900-905

[11] Ma W, Lu J, Wang B. Sliding friction and wear of Cu–graphite against 2024, AZ91D and Ti6Al4V at different speeds. Wear. 2009;**266**:1072-1081

[12] Dinaharan I, Saravanakumar S, Kalaiselvan K, Gopalakrishnan S. Microstructure and sliding wear characterization of Cu/TiB_2 copper matrix composites fabricated via friction stir processing. Journal of Asian Ceramic Societies. 2017;**5**(3):295-303

[13] Kumar J, Mondal S. Microstructure and properties of graphite-reinforced copper matrix composites. Journal of the Brazilian Society of Mechanical Sciences and Engineering. 2018;**40**:196

[14] Gautam RK, Ray S, Sharma SC, Jain SC, Tyagi R. Dry sliding wear behavior of hot forged and annealed Cu–Cr–graphite in-situ composites. Wear. 2011;**271**:658-664

[15] Nayak D, Ray N, Sahoo R, Debata M. Analysis of tribological performance of Cu hybrid composites reinforced with graphite and TiC using factorial techniques. Tribology Transactions. 2014;**57**:908-918

[16] Rajkumar K, Aravindan S. Tribological performance of microwave sintered copper–TiC–graphite hybrid composites. Tribology International. 2011;**44**:347-358

[17] Cullity BD. Elements of X-Ray Diffraction. 2nd ed. London, UK: Addision Wesley; 1978

[18] Singh MK, Gautam RK. Synthesis of copper metal matrix hybrid composites using stir casting technique and its mechanical, optical and electrical behaviours. Transactions of the Indian Institute of Metals. 2017;**70**:2415-2428. DOI: 10.1007/s12666-017-1103-0

[19] Rajkumar K, Aravindan S. Microwave sintering of copper–graphite composites. Journal of Materials Processing Technology. 2009;**209**:5601-5605

[20] Rajkovic V, Bozic D, Stasic J, Wang H, Jovanovic MT. Processing, characterization and properties of copper-based composites strengthened by low amount of alumina particles. Powder Technology. 2014;**268**:392-400

[21] Jha P, Gautam RK, Tyagi R. Friction and wear behavior of Cu–4 wt.%Ni–TiC composites under dry sliding conditions. Friction. 2017;**5**:437-446. DOI: 10.1007/s40544-017-0157-7

[22] Tu JP, Rong W, Guo SY, Yang YZ. Dry sliding wear behavior of in situ Cu–TiB_2 nanocomposites against medium carbon steel. Wear. 2003;**255**:832-835

[23] Kumar A, Jha PK, Mahapatra MM. Abrasive wear behavior of in situ TiC reinforced with Al-4.5%Cu matrix. Journal of Materials Engineering and Performance. 2014;**23**:743-752

[24] Prabhu TR, Varma VK, Vedantam S. Tribological and mechanical behavior of multilayer Cu/SiC + Gr hybrid composites for brake friction material applications. Wear. 2003;**255**:832-835

[25] Jha P, Gautam RK, Tyagi R, Kumar D. Sliding wear behavior of TiC-reinforced Cu-4 wt.% Ni matrix composites. Journal of Materials Engineering and Performance. 2016;**25**:1-9

[26] Sarmadi H, Kokabi AH, Reihani SMS. Friction and wear performance of copper–graphite surface composites fabricated by friction stir processing (FSP). Wear. 2013;**304**(1):1-12

[27] Saka N, Teixeira JJP, Suh NP. Wear of two phase metals. Wear. 1977;**44**:77-86

[28] Sundararajan S, Rajadurai JS. Investigation of microstructure, mechanical, and tribological properties of solid self-lubricating copper hybrid composites processed by solid-state mixing technique. Proceedings of the Institution of Mechanical Engineers, Part J: Journal of Engineering Tribology. 2016;**230**:40-56

Chapter 5

Surface Tension and Surface Tension Assessment of Ag-Au-Cu Ternary and Sub-Binary Alloy Systems

Hüseyin Arslan and Ali Dogan

Abstract

A brief review of measurement techniques and theoretical studies on the surface tension alloy and mixture has been presented in the present study. It is clear that the experimental determination of thermodynamic and thermophysical properties of both solid and especially liquid alloys at high temperature cases is frequently difficult technologically. In addition to this, a lack of experimental data concerning thermophysical properties of Ag-Au, Au-Cu, and Ag-Cu sub-binary systems is obvious. The theoretical thermophysical data of the Ag-Au-Cu ternary alloy systems are very scarce in the literature. Thus, the surface tensions of the alloys just mentioned above for cross sections $z = x_{Ag}/x_{Au} = 1/3, 1/1, 3/1, 2/5$, and 5/2, respectively, and their sub-binary systems are much simply calculated from the surface tensions of the Ag-Au, Au-Cu, and Ag-Cu sub-binary systems by using geometric models, such as Muggianu, Kohler, Toop, and GSM (Chou's general solution model) and Butler's equation. The predicted results in the present study show rather an agreement with the experimental results of the alloys. Therefore, it is inferred that the obtained surface tension curves for the Ag-Au-Cu ternary alloy at 1381 K are reasonable with especially those calculated from the Toop model.

Keywords: surface tension, geometric models, Butler equation, Pb-free ternary alloys

1. Introduction

It is seen that the surface tensions of the materials are of outstanding importance from many scientific and technological viewpoints. Surface tensions have been measured for a long time and it is seen that the collections of experimental data for pure liquids [1–11] and some binary liquid alloy systems exist. Surface tension measuring techniques can be classified generally as goniometric and tensiometric. They can also be classified into two classes. The first class is static surface tension measurements. When the values of surface tension are constant, the pure liquids are measured with these devices. The second one is dynamic surface tension measurements so that many of these are considered as the modifications of the static models. One can generally mention some experimental methods such as ring [1, 12]; oscillating jet [13–17]; DC method, as described in detail elsewhere [18–20]; oscillating droplet method [21–40]; draining crucible method [19, 20, 41, 42]; drop or weight method (it can be seen that the drop volume or weight method among the conventional methods of surface tension measurement has proven to be reliable and easy to handle) [43–47]; pulsating bubble [48]; pendant drop (it may be said that the use of the pendant drop method to measure interfacial tension between molten polymers has gotten a lot of attention) [11, 49–53]; sessile drop [54–78]; and maximum bubble pressure methods (this method is one of the most popular techniques to measure the dynamic surface tension of various surfactants).

This method is particularly useful in measuring surface tension of highly concentrated surfactant solutions [79] and molten metals [80–83] for binary alloys [84–90] for multicomponent alloys like Au-based [87, 91–106]. The last one among these methods has received much attention recently.

Recently, some researchers [19, 20, 41] developed a new method for fluids to simultaneously measure the surface tension, viscosity, and density in only one experiment.

Although a brief review of the experiments is given above, it is impossible to say that the experiments carried out are sufficient. On the other hand, the surface tension prediction is useful in designing and discovering new materials and it is necessary to discuss them theoretically. A brief review of some theoretical studies can be given here [16, 33, 71, 101, 104, 107–159] along with neural network modeling dealing with the alloys and mixtures. The artificial neural network (ANN) studies have been carried out to predict the surface tension of some chemicals including liquid drugs [160] and the alloys Sn-In-Zn-Ag using Butler model, rare earth containing binary chloride mixtures via STCBE computer program [161], and the binary alloys Fe-Cu, Cu-Pb, Sn-Pb, Ag-Pb, Pb-In, Bi-Ag, Ag-Sn, Cu-A1, Fe-Si, and Ni-Si via a special calculation technique [135, 162]. In addition, the surface tensions of the binary alloys in some low melting metal systems (Pb-Sb, Ag-Bi, Ag-In, Ag-Sn, Bi-Sn, and Sb-Sn) using a thermodynamic database and a database of densities and surface tensions of the respective pure metals based on published experimental data are given in ref. [163].

It is impossible to avoid traces of oxygen in the working atmosphere. In such situations, the surface contamination phenomena continues to be important in determining reliable thermophysical properties, such as data of surface tension, density, etc. Recently, for high-temperature measurements and studies of reactive materials, the containerless processing techniques in the experimental studies carried out have been used in order to isolate samples from their environment and to reduce contamination [164]. The surface tensions of liquid metals, Zr, Ni, Ti, Mo, and Nb, have been measured at their melting points using the quasi-containerless pendant drop method, which is much reliable for contamination [3]. Some critical investigations revealed that only a small number of papers concerning systematic investigations of surface tension of the multicomponent systems have been published on the effect of a wide range of temperature and concentration on the thermophysical properties. It can also be seen that the effect of a wide range of temperature and concentration on the surface tension of liquid ternary alloy systems is very sparse. Recently, the surface tension data for the Ag-Au, Au-Cu, and Ag-Cu binary alloys have been measured and reported in literature by some authors [165–169]. However, to the best of our knowledge, there is disagreement between data from different authors concerning the existence with agreements in the selected surface tensions of the components Ag, Au, and Cu. When an analytical expression of γ_i, experimental data, the surface tensions of the Ag-Au [90], Au-Cu [170], and Ag-Cu [59] binary alloys at 1381 K by using Muggianu model [171] were calculated and those of the Ag-Au-Cu ternary alloy at 1381 K by using geometric models, such as Kohler [172], Toop [173], GSM [174], and Butler's equation [175], were compared with the experimental data measured by Pajarre et al. [167] and Gallois and Lupis [166] in the present study. Hu et al. [168] first extended geometric models to predict the surface tensions of the Ag-Au-Cu ternary alloy but did not make a comparison between the calculated and experimental results. In the present study, the case in question was carried out and the ratios of silver/gold were kept constant, and the amount of copper was systematically changed with the ratio $z = x_{Ag}/x_{Au}$ = 1/3, 1/1, 3/1, 2/5, and 5/2, respectively while computing the surface tensions. Moreover, the symmetry properties of the Ag-Au-Cu ternary alloy system were judged. On the other hand, the excess surface tensions instead of excess thermodynamic properties encountered frequently in researches calculating the Gibbs energies and enthalpies of the alloys [176–183] have been used in order to calculate the surface tension of related ternary alloys in the present study [2, 46, 62, 86, 97, 100, 174, 176–181, 184–188].

2. Background of excess and ideal surface tension and geometric models

Excess surface tension (σ^E) has been calculated by Eq. (1):

$$\sigma^E = \sigma - \sigma^i \tag{1}$$

where

$$\sigma^i = \sum_{i=1}^{2} x_i \sigma_i \tag{2}$$

here, x_i and σ_i are the mole fraction and surface tension of the pure liquid component i.

A thermodynamic property of mixing, such as surface tension, can be expressed as a combination of all binaries with an assigned probability weight and the thermodynamic properties of a ternary system can be then calculated by

$$\sigma^E = \sum_{ij} W_{ij}\sigma^E_{ij}, \ (i \neq j = 1 \text{ to } 3) \tag{3}$$

where W_{ij} represents the weight probability of each corresponding binary composition point and can be calculated via:

$$W_{ij} = \frac{x_i x_j}{X_{i(ij)} X_{j(ij)}} \tag{4}$$

where $X_{i(ij)}$ and $X_{j(ij)}$ indicate the mole fractions of component i and j in the binary system and are given different forms according to the selected geometric models.

The excess surface tension in Eq. (3) associated with the binary subsystems of the alloy systems is given in the form of Redlich-Kister polynomials:

$$\sigma^E{}_{ij} = x_i x_j \left[\left(A^0_{ij} + B^0_{ij} T \right) + \left(A^1_{ij} + B^1_{ij} T \right) (x_i - x_j) + \left(A^2_{ij} + B^2_{ij} T \right) (x_i - x_j)^2 + \cdots \right] \tag{5}$$

The Muggianu model can be expressed readily in a series form so that A_{ij} denotes the Redlich-Kister parameters:

$$\sigma^E = \sum_{i=1}^{2} \sum_{j>i}^{3} x_i x_j A_{ij} + x_i\, x_{j>i}\, x_{k>j}\, A_{i,j,k} \tag{6}$$

where

$$A_{ij} = \sum_{\vartheta=0}^{2} A_{ij}^{\vartheta} \left(x_i - x_j\right)^{\vartheta} \tag{7}$$

$$A_{i,j,k} = x_i A_{i,j,k}^{0} + x_j A_{i,j,k}^{1} + x_k A_{i,j,k}^{2} \tag{8}$$

The Kohler model can be given in a closed form simply:

$$\begin{aligned}\sigma^E = {} & (x_1 + x_2)^2 \sigma^E{}_{12} \left(\frac{x_1}{x_1 + x_2}; \frac{x_2}{x_1 + x_2} \right) + (x_3 + x_1)^2 \sigma^E{}_{31} \left(\frac{x_3}{x_1 + x_3}; \frac{x_1}{x_1 + x_3} \right) \\ & + (x_2 + x_3)^2 \sigma^E{}_{23} \left(\frac{x_2}{x_2 + x_3}; \frac{x_3}{x_2 + x_3} \right)\end{aligned} \tag{9}$$

The Toop model can also be given in a closed form simply:

$$\sigma^E = \frac{x_2}{1 - x_1} \sigma^E{}_{12}(x_1; 1 - x_1) + \frac{x_3}{1 - x_1} \sigma^E{}_{13}(x_1; 1 - x_1) + (x_2 + x_3)^2 \sigma^E{}_{23} \left(\frac{x_2}{x_2 + x_3}; \frac{x_3}{x_2 + x_3} \right) \tag{10}$$

Using Eqs. (3) and (4), the expression of the surface tension concerning the GSM model can be written as:

$$\sigma^E = \frac{X_1 X_2}{X_{1(12)} X_{2(12)}} \sigma^E{}_{12} + \frac{X_2 X_3}{X_{2(23)} X_{3(23)}} \sigma^E{}_{23} + \frac{X_3 X_1}{X_{3(31)} X_{1(31)}} \sigma^E{}_{31} \tag{11}$$

This model gets rid of the asymmetric component selecting problems seen in the Toop model. The compositions of binary alloys in Eq. (11) can be written as:

$$X_{1(12)} = X_1 + X_3 \xi_{12} \tag{12}$$

$$X_{2(23)} = X_2 + X_1 \xi_{23} \tag{13}$$

$$X_{3(31)} = X_3 + X_2 \xi_{31} \tag{14}$$

The key step in the GSM is introducing a new concept, called a similarity coefficient, ξ, into the geometrical model, that is,

$$\xi_{ij}^{k} = \frac{\eta(ij, ik)}{\eta(ij, ik) + \eta(ji, jk)} \tag{15}$$

where $\eta(ij,\, ik)$ is a function related to the excess Gibbs free energy of ij and ik two binaries and is given as follows:

$$\begin{aligned}\eta(ij, ik) = {} & \int_{x_i - 0}^{x_i - 1} \left(\sigma_{ij}^{E} - \sigma_{ik}^{E} \right)^2 dX_i = \eta(ij, ik) = \sum_{l=0}^{n} \frac{1}{2(2l+1)(2l+3)(2l+5)} \left(A_{ij}^{l} - A_{ik}^{l} \right)^2 \\ & + \sum_{l=0}^{n} \sum_{m>l}^{n} \frac{1}{(l+m+1)(l+m+3)(l+m+5)} \times \left(A_{ij}^{l} - A_{ik}^{l} \right) \left(A_{ij}^{m} - A_{ik}^{m} \right)\end{aligned} \tag{16}$$

Its value up to three order of k, for instance, can be given as follows readily:

$$\eta(12,13) = \frac{1}{30}\left(A_{12}^{0} - A_{13}^{0}\right)^2 + \frac{1}{210}\left(A_{12}^{1} - A_{13}^{1}\right)^2 + \frac{1}{630}\left(A_{12}^{2} - A_{13}^{2}\right)^2 + \frac{1}{105}\left(A_{12}^{0} - A_{13}^{0}\right)\left(A_{12}^{2} - A_{13}^{2}\right) \tag{17}$$

In addition, it should be noted that $A_{ij}^{k} = A_{ji}^{k}$, when k is even and $A_{ij}^{k} = -A_{ji}^{k}$ and when k is odd in Eqs. (6), (7), (16), and (17).

3. Background of the Butler model

Butler's equation has been used extensively to calculate the surface tension of multicomponent alloy systems. The surface of the bulk is treated as an additional thermodynamic phase, in equilibrium with the bulk phase. If the ternary alloys are taken as a regular solution model, the surface tension of ternary liquid alloys dealing with Butler's model is written as:

$$\begin{aligned} \sigma &= \left\{\sigma_1 + \frac{RT}{S_1}\ln\left(\frac{X_1^s Y_1^s}{X_1^b Y_1^b}\right)\right\} \\ \sigma &= \left\{\sigma_2 + \frac{RT}{S_2}\ln\left(\frac{X_2^s Y_2^s}{X_2^b Y_2^b}\right)\right\} \\ \sigma &= \left\{\sigma_3 + \frac{RT}{S_3}\ln\left(\frac{X_3^s Y_3^s}{X_3^b Y_3^b}\right)\right\} \end{aligned} \tag{18}$$

here, R, T, σ_i, and S_i are gas constant, temperature in terms of K, surface tension of pure component i, and surface area of component i, respectively. The surface area of component i is

$$S_i = 1.091 N_a^{1/3}\left(\frac{M_i}{\rho_i}\right)^{2/3} \tag{19}$$

here, the molar volume of each component V_i can be expressed as M_i/ρ_i. This expression can be calculated from Avogadro's number N_a, the atomic mass M_i, and the density data ρ_i, as in Eq. (19). In this equation, X_j^k represents an alloy composition with the subscript j and superscript k referring to the corresponding component j in the bulk, b, and the surface phase, s. The terms γ_i^b (T, X_j^b) and γ_i^s (T, X_j^s) in Eq. (18) are activity of component i in the bulk phase and the surface phase, respectively. These terms can be obtained as both functions of temperature and composition. The activity term of component i can be derived from standard thermodynamic relationships, in the form:

$$G_i^E = RT \ln\, \gamma_i = G^E + \sum_{j=1}^{3}\left(\delta_{ij} - x_j\right)\frac{\partial G^E}{\partial X_j} \tag{20}$$

where δ_{ij} is Kronecker's symbol and j = i, $\delta_{ij} = 1$ and $j \neq i$, $\delta_{ij} = 0$. Tanaka et al. [184] have proposed a model for $G_i^{E,s}$ (T, $X^s_{j\,(j=2;3,\ldots)}$) as follows:

$$G_i^{E,s}\left(T,X_j^s\right) = \beta\, G_i^{E,b}\left(T,X_j^b\right) \tag{21}$$

where β is a parameter corresponding to the ratio of the coordination number z in the surface phase to that in the bulk phase, z^s/z^b. In some cases, the value of β parameter might be affected by other factors, such as the relaxation of the surface structure, and in the literature, different values between 0.5 and 0.84 are used [62]. The number of nearest neighbors surrounding a central atom is 9 in the surface phase. In addition, the coordination number for the bulk atom is 12. Here, in order to calculate surface tensions of the alloy systems, the value of the parameter β has been taken as 0.83 in the present study.

In order to measure surface tension of the liquids at temperatures above the melting point of alloys on the one hand, the surface tensions of some alloys have been measured contactless using the technique of electromagnetic levitation in combination with the oscillating drop technique [127]. On the other hand, some calculation models have also been developed. In order to calculate surface tension of the binary alloys [170], a new thermodynamic model is proposed by Prasad and Mikula [143]. It is assumed that there is a relation between γ_i,γ_i^s and p ln γ_i^* , q lnγ_i; here, p and q are called as surface coordination fractions in which p = 0.5, q = 0.25 for closed packed structured. When an analytical expression of γ_i is known, it is possible to obtain expressions for γ_i^s, so the surface tension of the binary mixture can be calculated easily. Recently, the surface properties of Ag-Cu and Ag-Ti liquid alloys are predicted by a quasi-chemical solution model (QCS) [189–192]. Whileas the compound formation model (CFM) [162, 193] (it is related with both weak and strong interaction approximation) in Cu-Ti system has been performed. Moreover, the surface tensions of Ag-Ti and Ag-Hf liquid alloys have been predicted by the QCA in the case of regular solutions, developed by Bhatia and Singh [194, 195], in the framework of statistical mechanical theory in conjunction with the quasi-lattice theory, while in the case of the Ag-Ti system, at T = 1773, the CFM has been applied [190, 196].

On the other hand, the value of coefficient f = 1.091 in the surface area of component i in Eq. (19) has been altered by Kaptay and Papp [197]. In addition to this, the model of Butler has been reconstructed from the very beginning to finishing of the model for the surface tension of one-component liquid metals. In their paper, the effects of surface active complexes such as the intermetallic compounds in liquid metallic solutions are also considered.

4. Experimental surface tension data of liquid Ag, Au, and Cu

The surface tensions of liquid Ag, Au, and Cu were measured at 1381 K by Gallois and Lupis [166] under atmospheres of argon, dilute solutions of hydrogen in argon, and under vacuum and are found as σ_{Ag} = 0.890 ± 0.01 (N/m), σ_{Au} = 1.150 ± 0.01 (N/m), and σ_{Cu} = 1.320 ± 0.015 (N/m). Pajarre et al. [167] have used the values σ_{Ag} = 0.892132 [46], σ_{Au} = 1.13666 (N/m), respectively [185], and σ_{Cu} = 1.29499 (N/m) [21] in their experimental and theoretical study in which the experimental surface tension values for the Ag-Au-Cu system have been determined by the sessile drop method at 1381 K [167]. For the surface tension calculations in Surdat, the surface tension data σ_{Ag} = 0.870912 (N/m) [186], σ_{Au} = 1.503 (N/m) [2], and σ_{Cu} = 1.4756 (N/m) [100] are used. As mentioned previously, the surface tension data are very dispersed in the literature. For the calculations carried out in the present study, the values of the surface tension of [166] have been adopted since it is inferred that the experimental results deviate from the corresponding values predicted by the geometric models at the temperature of 1381 K.

System	A_{ij^0}	A_{ij^1}	A_{ij^2}
Ag-Au	−128.79 + 0.045836 T	243.86 − 0.060170 T	−455.82 + 0.185348 T
	+0.00680219 T lnT	−0.00965926 T lnT	+0.02786783 T lnT
	−0.00001833 T^2	+0.00002163 T^2	−0.00011934 T^2
Ag-Cu	−982.68 + 0.262677 T	1625.72 − 0.693760 T	−1142.64 − 0.084365 T
	+0.04028137 T lnT	−0.09772514 T lnT	+0.14883052 T lnT
	−0.00013536 T^2	+ 0.00039080 T^2	−0.00030951 T^2
Au-Cu	94.97 + 0.002865 T	−43.72−0.009243 T	−9.72 − 0.008609 T
	−0.00050101 T lnT	−0.00094336 T lnT	−0.00071243 T lnT
	−0.0000217 T^2	+ 0.00000558 T^2	+ 0.00002569 T^2

Table 1.
Redlich-Kister parameters for excess surface tensions of the Ag-Au, Ag-Cu, and Au-Cu sub-binary alloys at 1381 K.

By using the values of Redlich-Kister parameters of the binary alloys given in references [2, 86, 97] and the surface tensions of liquid Ag, Au, and Cu measured at 1381 K in the reference [166], the excess binary surface tensions σE and binary surface tensions σ of three sub-binary systems are calculated and shown in **Figure 1**. Using Eqs. (1)–(4) and the other equations relevant to the models, the surface tension reference data for the pure components, such as the surface tension of

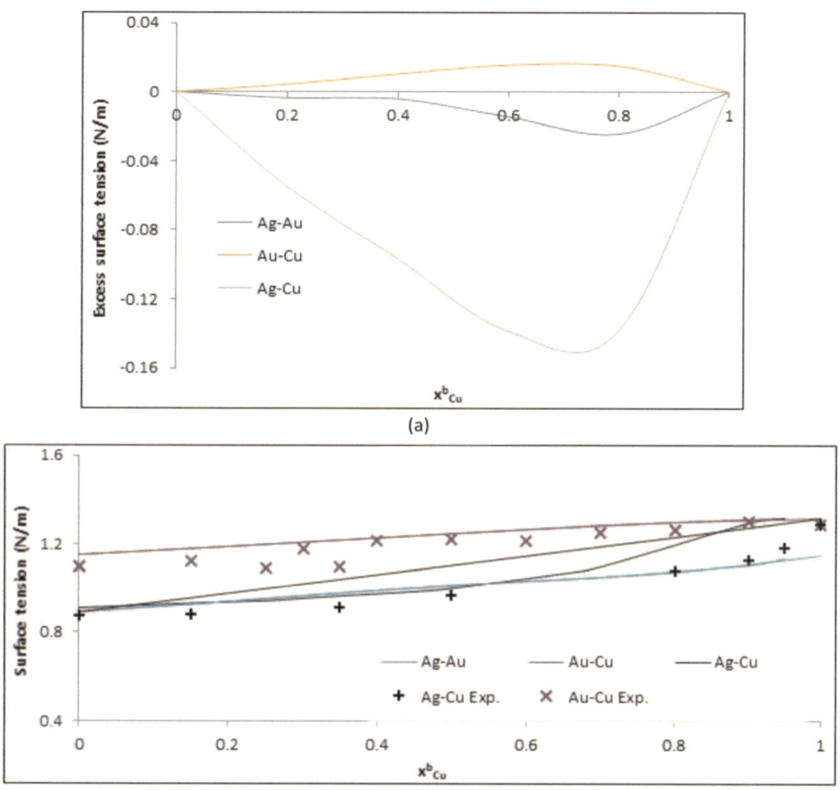

(b)

Figure 1.
(a) The excess binary surface tensions σ^E of three sub-binary systems in the Ag-Au-Cu ternary alloy at 1381 K and (b) the binary surface tensions σ of three sub-binary systems in the Ag-Au-Cu ternary alloy at 1381 K.

V_{Au} (m^3/mol) = 11.3 (1 + 0.8/10,000 (T/K – 1337.33))/1,000,000	[185]
V_{Ag} (m^3/mol) = 11.6 (1 + 0.98/10,000 (T/K – 1234.93))/1,000,000	[46]
V_{Cu} (m^3/mol) = 7.94 (1 + 1/10,000 (T/K – 1357.77))/1,000,000	[46]

Table 2.
Molar volume data for the Ag-Au-Cu alloy system.

System	A_{ij^0}	A_{ij^1}	A_{ij^2}
Ag-,Au	−16,402 + 1.14 T [198]	—	—
Ag-,Cu	17384.37 – 4.46438 T [199]	1660.74 – 2.31516 T [199]	—
Au-,Cu	−27,900 – T [200]	4730 [200]	3500 + 3.5 T [200]
Ag-,Au-,Cu	10,000 [199]	105,000 + 30 T [199]	−1000 [199]

Table 3.
Interaction parameters for excess Gibbs energy in the Ag-Au-Cu system.

liquid Ag, Bi, Sn, and the molar volumes (**Table 2**) [46, 185], were combined with the Gibbs energies (in order to calculate these energies, Redlich-Kister parameters given in **Table 3** are used) of the binary and ternary systems in order to calculate the surface tension of the liquid Ag-Au-Cu alloy systems using the Butler model.

5. Results and discussion

Geometrical models have been recently successfully applied to predict the surface tension of the Ni_3S_2-FeS-Cu_2S system, and the obtained results were in good agreement with those of experimental values [201]. Moreover, the surface tensions of the Sn-Ga-In ternary alloy systems were calculated from the surface tensions of the Sn-Ga, Ga-In and In-Sn sub-binary systems at 773 K using geometric models, such as the Kohler, Toop, and GSM, by some researchers [140]. In the Toop model, which is a classic asymmetric geometric model, it is very important to determine the asymmetric component [202]. As is expressed in previous papers [176–183, 199], if the deviations of the binary systems A-B and A-C from the ideal solution are similar, but differ markedly from those of the binary system B-C, then the A-B-C ternary system is asymmetric, and the ternary system mentioned above becomes symmetric. It can be readily said that the common component A in two sub-binary systems with thermodynamic similarities ought to be chosen as the asymmetric component in the asymmetric system in the Toop model. When the excess binary surface tensions σ^E of three sub-binary systems in the Ag-Au-Cu ternary alloy at 1381 K are compared, it can clearly be seen that the asymmetric component is Au because the two binary Ag-Au and Au-Cu systems are much more similar thermodynamically, as shown in **Figure 1**. The similarity coefficients for the calculated excess surface tension of the Ag-Au, Ag-Cu, and Au-Cu sub-binary alloys mentioned in GSM have been calculated and found as 0.934915, 0.600985, and 0.044179 (**Table 4**), respectively. On glancing at the values of the similarity coefficients, it is seen that none of the three similarity coefficients is approximately equivalent to unity. So, the asymmetric component in the ternary alloys is not easy to find. Here, ξ_{Ag-Au} = 0.82 and this means that Ag is relatively similar to Cu. It also indicates that someone has no alternative option but to

Deviation of sum of squares	$\eta_1 = 8051.52$		$\eta_2 = 560.51$		$\eta_3 = 12126.95$	
Binary systems	Ag-Au	Ag-Cu	Au-Ag	Au-Cu	Cu-Ag	Cu-Au
Similarity coefficients	$\xi_{Ag\text{-}Au} = 0.934915$		$\xi_{Au\text{-}Cu} = 0.044179$		$\xi_{Cu\text{-}Ag} = 0.600985$	

Table 4.
Deviation of sum of squares and similarity coefficients associated with surface tension in Ag-Au-Cu ternary system for GSM at 1381 K.

select Au as the asymmetric component. By investigating the abovementioned analysis, it can be concluded that the Ag-Au-Cu ternary system is not equivalent to the Muggianu, the Kohler, or the Toop models, so both the Muggianu model and the Kohler model cannot obtain the predicted values accurately. Therefore, the predicted surface tensions calculated by the GSM model can be recommended. In addition, according to the abovementioned analysis, one can conclude that the Ag-Au-Cu ternary system is exactly not determined by the symmetric models. A comparison of the surface tension values, which are calculated by the selected models, with the experimental values measured in literature [166, 167] for cross sections (**Figure 4**) (a) $z = x_{Ag}/x_{Au} = 1/3$, (b) 1/1, (c) 3/1, (d) 2/5, and (e) 5/2, in the Ag-Au-Cu ternary system is shown in **Figure 2**. In addition, it is inferred from the discussion of the standard errors (**Table 5**) of the calculation results performed in the present study that the best model is the Toop model for all cross sections $z = x_{Ag}/x_{Au} = 1/3$, 1/1, 3/1, 2/5, and 5/2, in the whole experimental range. Therefore, the excess surface tension curves of the Ag-Au, Ag-Cu, and Au-Cu sub-binaries calculated by Eq. (5) and Ag-Au-Cu ternary alloy systems at 1381 K by the Toop model are plotted in **Figure 3**, since the Toop model is found as the best model compared to the other models. For example, the comparison between the calculated surface tension for cross section $x_{Ag}/x_{Au} = 1/3$ in the Ag-Au-Cu alloy system and its experimental values at 1381 K is provided in **Table 6**.

On the other hand, a comparison of the surface tension values calculated via the Toop model with the experimental values measured in literature [167] for cross section $z = x_{Ag}/x_{Au} = 1/3$, 1/1, 3/1, 2/5, and 5/2 in the Ag-Au-Cu ternary systemis given in **Figure 3a**. The Toop model is able to describe the experimental data for all ratios mentioned above. In addition to this, the dependency of the surface molar fraction of copper from bulk composition in the Ag-Au-Cu ternary system is calculated by the Butler equation and its plot is given in **Figure 3b**. One can obtain the distribution between interface region and bulk phase for all five components in Ag-Au-Cu ternary mixture calculating the surface mole fraction $x_{i,s}$ for a given bulk-phase mole fraction *xi*, B via the Butler equations (Eq. (18)). When the ratio of x_i,s/*xi*,b is equal to one, it yields no surface active enrichment/depletion owing to a reference line. At constant ratios $z = x_{Ag}/x_{Au} = 1/3$, 1/1, 3/1, 2/5, and 5/2, the relative enrichment of all three components are depicted in **Figure 3b**. It can be seen at the relative high or low concentration x_{Ag}/x_{Au}, in the alloys mentioned above, that concentration of Cu can be found in the bulk phase and show no tendency to enter the surface region. It can be also seen from **Figure 3a** that the most striking aspects of the surface tension calculations carried out in the present study for all models can improve the surface tension values when Cu is added to the alloys with Ag and Cu contents.

A comparison of the surface tension values calculated for binary Ag-Au, Ag-Cu, and Au-Cu alloys with the experimental values measured in literature except for Ag-Au [167] is given in **Figure 1b**. On the other hand, it is calculated by Calvo [169] that the surface tensions of Ag-Au binary alloys are found between 550 and 700 mN/m at a temperature of 1500 K. These results are also in agreement with those obtained experimentally by ref. [57], and those calculated in this study.

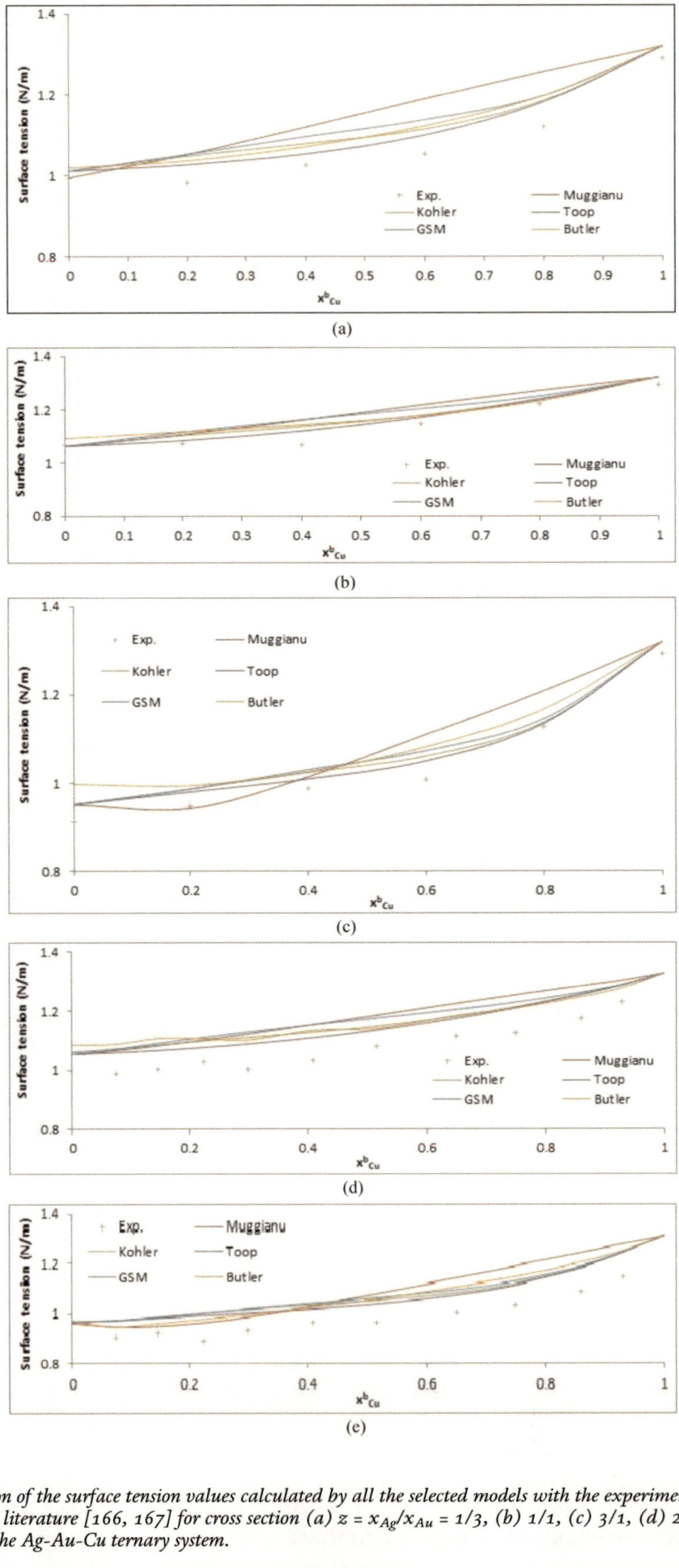

Figure 2.
A comparison of the surface tension values calculated by all the selected models with the experimental values measured in literature [166, 167] for cross section (a) $z = x_{Ag}/x_{Au} = 1/3$*, (b) 1/1, (c) 3/1, (d) 2/5 and (e) 5/2, in the Ag-Au-Cu ternary system.*

Sections	Muggianu	Kohler	Toop	GSM	Butler
Ag/Au = 1/1	0.038259	0.021469	0.016782	0.026046	0.021911
Ag/Au = 1/3	0.022055	0.014311	0.010802	0.020502	0.016629
Ag/Au = 3/1	0.023755	0.015022	0.012579	0.016889	0.022797
Ag/Au = 2/5	0.029510	0.023499	0.019233	0.028573	0.024484
Ag/Au = 5/2	0.028877	0.024461	0.021560	0.026221	0.024006

Table 5.
The standard errors associated with surface tension in Ag-Au-Cu ternary system.

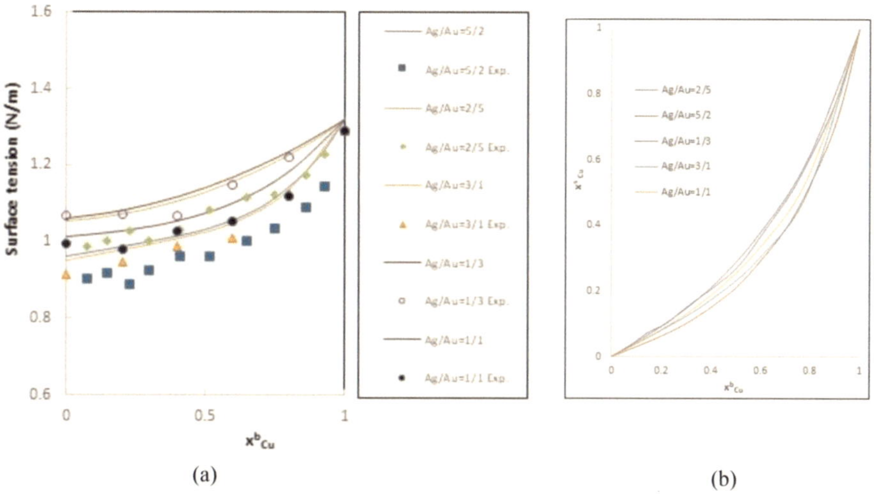

Figure 3.
(a) A comparison of the surface tension values calculated by Toop model with the experimental values measured in literature [166, 167] for all cross sections ($z = x_{Ag}/x_{Au}$ = 1/3, 1/1, 3/1, 2/5, and 5/2) in this figure concerning the Ag-Au-Cu ternary system. (b) Dependency of the surface-bulk composition of copper in the Ag-Au-Cu ternary system for all cross sections ($z = x_{Ag}/x_{Au}$ = 1/3, 1/1, 3/1, 2/5, and 5/2).

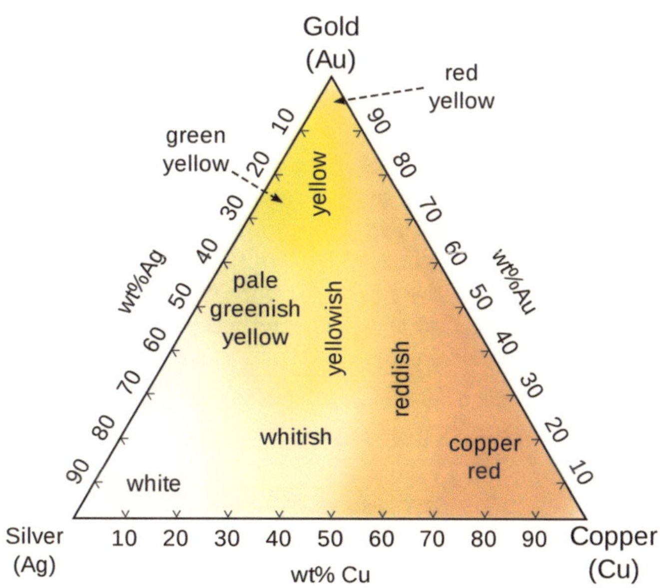

Figure 4.
A phase diagram of the Ag-Au-Cu ternary alloy [203].

x^b_{Cu}	Muggianu	Kohler	Toop	GSM	Butler	Exp.
0	1.0620	1.0620	1.0620	1.0620	1.0922	1.067
0.2	1.1051	1.1018	1.0828	1.1079	1.1163	1.071
0.4	1.1593	1.1346	1.1181	1.1474	1.1423	1.069
0.6	1.2160	1.1770	1.1692	1.1893	1.1767	1.147
0.8	1.2701	1.2382	1.2365	1.2433	1.2307	1.220
1	1.3200	1.3200	1.3200	1.3200	1.3200	1.291

Table 6.
A comparison between the calculated surface tension for cross section x_{Ag}/x_{Au} = 1/3 in the Ag-Au-Cu alloy system and its experimental values at 1381 K [167].

Taking into consideration the fact that there is no sufficient experimental thermophysical data in literature, it is envisaged that the results obtained from the theoretical analysis of the alloy systems in the present work might fill the research gap on the thermophysical properties such as surface tension. Therefore, there are no agreements among the models except for the cross sections $z = x_{Ag}/x_{Au} = 1/1, 3/1$. According to the researchers, the high temperature might be one of the factors that mainly affect the surface tension so that the oxidation at high temperatures directly affects surface tension and will give rise to decreased values.

The major experimental measuring difficulty in surface tension measurements of the multicomponent alloys at high temperatures lies in the control of the processing atmosphere existing in device medium. Taking into account the difficulty in measuring the surface tension of multicomponent alloys, surface tension of the multicomponent alloys was calculated by calculating the surface tension of two-component liquid alloys and demonstrated their general applicability in this study.

6. Conclusions

For cross sections $z = x_{Ag}/x_{Au} = 1/3, 1/1, 3/1, 2/5$, and 5/2, the surface tension values of the treated alloys Ag-Au-Cu have been modeled and analyzed theoretically by using such approaches to geometric models, as the Muggianu, the Kohler, the Toop, and the GSM models, and Butler's equation. Considering that the predictions of all models are generally consistent with the experimental findings, some conclusions are given as follows: the surface tension values increase as the composition of Cu increases. It is seen at the relative high or low concentration x_{Ag}/x_{Au} in the above-mentioned alloys that concentration of Cu can be found in the bulk phase and shows no tendency to enter the surface region. It is inferred from the calculations performed for all models that the Toop method yields in a great promise to efficiently determine the physical properties such as surface tension data of alloys Ag-Au-Cu.

Author details

Hüseyin Arslan[1,2]* and Ali Dogan[1]

1 Department of Physics, Science and Arts Faculty, Kahramanmaras Sutcuimam University, Onikisubat, Kahramanmaras, Turkey

2 Department of Material Science and Engineering, Institute for Graduate Studies in Science and Technology, Kahramanmaras Sutcuimam University, Onikisubat, Kahramanmaras, Turkey

*Address all correspondence to: hseyin_arslan@yahoo.com

References

[1] Keene B. Review of data for the surface tension of pure metals. International Materials Reviews. 1993; **38**:157-192

[2] Gale W, Totemeier T, editors. Smithells Metals Reference Book. 8th ed. Amsterdam, Amsterdam Boston, Heidelberg, London, New York, Oxford, Paris, San Diego, San Francisco, Singapore, Sydney, Tokyo: Elsevier; 2004

[3] Man KF. Surface tension measurements of liquid metals by the quasi-containerless pendant drop method. International Journal of Thermophysics. 2000;**21**:793-804

[4] Egry I, Lohoefer G, Jacobs G. Surface tension of liquid metals: Results from measurements on ground and in space. Physical Review Letters. 1995;**75**:4043

[5] Egry I, Sauerland S. Containerless processing of undercooled melts: Measurements of surface tension and viscosity. Materials Science and Engineering A. 1994;**178**:73-76

[6] Egry I, Brillo Jr. Surface tension and density of liquid metallic alloys measured by electromagnetic levitation. Journal of Chemical & Engineering Data. 2009;**54**:2347-2352

[7] Lee J, Kiyose A, Nakatsuka S, Nakamoto M, Tanaka T. Improvements in surface tension measurements of liquid metals having low capillary constants by the constrained drop method. ISIJ International. 2004;**44**:1793-1799

[8] Ricci E, Giuranno D, Sobczak N. Further development of testing procedures for high temperature surface tension measurements. Journal of Materials Engineering and Performance. 2013;**22**:3381-3388

[9] Falke W, Schwaneke A, Nash R. Surface tension of zinc: The positive temperature coefficient. Metallurgical Transactions B. 1977;**8**:301-303

[10] Brillo J. Thermophysical Properties of Multicomponent Liquid Alloys. Berlin, Boston: Walter de Gruyter GmbH & Co KG; 2016

[11] Amore S, Giuranno D, Novakovic R, Ricci E, Nowak R, Sobczak N. Thermodynamic and surface properties of liquid Ge-Si alloys. Calphad. 2014;**44**: 95-101

[12] Thiessen D, Man K. Surface tension measurement. In: The Measurement, Instrumentation and Sensors Handbook on CD-ROM. Boca Raton: CRC Press; 1999

[13] Defay R, Hommelen JR. I. Measurement of dynamic surface tensions of aqueous solutions by the oscillating jet method. Journal of Colloid Science. 1958;**13**:553-564

[14] Defay R, Pétré G. Correcting surface tension data obtained by the oscillating jet method. Journal of Colloid Science. 1962;**17**:565-569

[15] Bellizia G, Megaridis CM, McNallan M, Wallace DB. A capillary-jet instability method for measuring dynamic surface tension of liquid metals. In: Proceedings of the Royal Society of London A: Mathematical, Physical and Engineering Sciences. The Royal Society; 2003. pp. 2195-2214

[16] Dubberstein T, Heller H-P, Klostermann J, Schwarze R, Brillo J. Surface tension and density data for Fe-Cr-Mo, Fe-Cr-Ni, and Fe-Cr-Mn-Ni steels. Journal of Materials Science. 2015;**50**:7227-7237

[17] Howell E, Megaridis C, McNallan M. Dynamic surface tension measurements of molten Sn/Pb solder using oscillating slender elliptical jets. International

Journal of Heat and Fluid Flow. 2004;**25**: 91-102

[18] Gancarz T, Moser Z, Gąsior W, Pstruś J, Henein H. A comparison of surface tension, viscosity, and density of Sn and Sn-Ag alloys using different measurement techniques. International Journal of Thermophysics. 2011;**32**: 1210-1233

[19] Roach S, Henein H. A dynamic approach to determining the surface tension of a fluid. Canadian Metallurgical Quarterly. 2003;**42**: 175-186

[20] Roach SJ, Henein H. A new method to dynamically measure the surface tension, viscosity, and density of melts. Metallurgical and Materials Transactions B. 2005;**36**:667-676

[21] Stückrad B, Hiller W, Kowalewski T. Measurement of dynamic surface tension by the oscillating droplet method. Experiments in Fluids. 1993;**15**: 332-340

[22] Cao C, Zhang L, Bai X, Duan L, Wang F. Measurement of surface tension and specific heat of Ni-18.8 at.% Si alloy melt by containerless processing. Journal of Materials Science. 2011;**46**:6243-6247

[23] Schmitz J, Brillo J, Egry I, Schmid-Fetzer R. Surface tension of liquid Al-Cu binary alloys. International Journal of Materials Research. 2009;**100**:1529-1535

[24] Dai F, Cao C, Wei B. Thermophysical properties of Ni-5% Sn alloy melt. Science in China Series G: Physics Mechanics and Astronomy. 2006;**49**:236-245

[25] Matsumoto T, Fujii H, Ueda T, Kamai M, Nogi K. Measurement of surface tension of molten copper using the free-fall oscillating drop method. Measurement Science and Technology. 2005;**16**:432

[26] Wang H, Cao C, Wei B. Thermophysical properties of a highly superheated and undercooled Ni-Si alloy melt. Applied Physics Letters. 2004;**84**: 4062-4064

[27] Egry I, Giffard H, Schneider S. The oscillating drop technique revisited. Measurement Science and Technology. 2005;**16**:426

[28] Kobatake H, Brillo J, Schmitz J, Pichon P-Y. Surface tension of binary Al-Si liquid alloys. Journal of Materials Science. 2015;**50**:3351-3360

[29] Ravera F, Loglio G, Kovalchuk VI. Interfacial dilational rheology by oscillating bubble/drop methods. Current Opinion in Colloid & Interface Science. 2010;**15**:217-228

[30] Seyhan I, Egry I. The surface tension of undercooled binary iron and nickel alloys and the effect of oxygen on the surface tension of Fe and Ni. International Journal of Thermophysics. 1999;**20**:1017-1028

[31] Schneider S, Egry I, Seyhan I. Measurement of the surface tension of undercooled liquid $Ti_{90}Al_6V_4$ by the oscillating drop technique. International Journal of Thermophysics. 2002;**23**: 1241-1248

[32] Fujii H, Matsumoto T, Izutani S, Kiguchi S, Nogi K. Surface tension of molten silicon measured by microgravity oscillating drop method and improved sessile drop method. Acta Materialia. 2006;**54**:1221-1225

[33] Brillo J, Plevachuk Y, Egry I. Surface tension of liquid Al-Cu-Ag ternary alloys. Journal of Materials Science. 2010;**45**:5150-5157

[34] Egry I, Brillo J, Holland-Moritz D, Plevachuk Y. The surface tension of liquid aluminium-based alloys. Materials Science and Engineering A. 2008;**495**:14-18

[35] Schick M, Brillo J, Egry I. Thermophysical properties of liquid Co-Cu-Ni alloys. International Journal of Cast Metals Research. 2009;**22**:82-85

[36] Han X, Wei B. Thermophysical properties of undercooled liquid Co-Mo alloys. Philosophical Magazine. 2003;**83**: 1511-1532

[37] Brillo J, Egry I. Density and surface tension of electromagnetically levitated Cu-Co-Fe alloys. International Journal of Thermophysics. 2007;**28**:1004-1016

[38] Wang H, Luo B, Qin T, Chang J, Wei B. Surface tension of liquid ternary Fe-Cu-Mo alloys measured by electromagnetic levitation oscillating drop method. The Journal of Chemical Physics. 2008;**129**:124706

[39] Brillo J, Egry I. Surface tension of nickel, copper, iron and their binary alloys. Journal of Materials Science. 2005;**40**:2213-2216

[40] Chang J, Wang H, Zhou K, Wei B. Surface tension measurement of undercooled liquid Ni-based multicomponent alloys. Philosophical Magazine Letters. 2012;**92**:428-435

[41] Henein H. The transition from free stream flow to dripping in draining vessels. Canadian Metallurgical Quarterly. 2005;**44**:261-264

[42] Trybula M, Gancarz T, Gasior W, Pasturel A. Bulk and surface properties of liquid Al-Li and Li-Zn alloys. Metallurgical and Materials Transactions A. 2014;**45**:5517-5530

[43] Tate T. XXX. On the magnitude of a drop of liquid formed under different circumstances. The London, Edinburgh, and Dublin Philosophical Magazine and Journal of Science. 1864;**27**:176-180

[44] Harkins WD, Brown F. The determination of surface tension (free surface energy), and the weight of falling drops: The surface tension of water and benzene by the capillary height method. Journal of the American Chemical Society. 1919;**41**:499-524

[45] Park SH, Um YS, Hur BY. Rheological properties of molten Al-Cu alloys for manufacturing metallic foam. Solid State Phenomena. 2006;**116**: 656-660

[46] Lee J, Shimoda W, Tanaka T. Surface tension and its temperature coefficient of liquid Sn-X (X= Ag, Cu) alloys. Materials Transactions. 2004;**45**: 2864-2870

[47] Plevachuk Y, Sklyarchuk V, Gerbeth G, Eckert S, Novakovic R. Surface tension and density of liquid Bi-Pb, Bi-Sn and Bi-Pb-Sn eutectic alloys. Surface Science. 2011;**605**:1034-1042

[48] Chang C-H, Coltharp KA, Park SY, Franses EI. Surface tension measurements with the pulsating bubble method. Colloids and Surfaces A: Physicochemical and Engineering Aspects. 1996;**114**:185-197

[49] Arashiro EY, Demarquette NR. Use of the pendant drop method to measure interfacial tension between molten polymers. Materials Research. 1999;**2**: 23-32

[50] Kumar A, Lehnert-Batar S, Mäder R, Windhab EJ. Measurement of the surface and interfacial tension from maximum volume of a pendant drop. Journal of Colloid and Interface Science. 2001;**244**:113-122

[51] Semmler A, Kohler H-H. Surface properties of alkylpyridinium chlorides and the applicability of the pendant drop technique. Journal of Colloid and Interface Science. 1999;**218**:137-144

[52] Andreas J, Hauser E, Tucker W. Boundary tension by pendant drops 1. The Journal of Physical Chemistry. 1938; **42**:1001-1019

[53] Stauffer CE. The measurement of surface tension by the pendant drop technique. The Journal of Physical Chemistry. 1965;**69**:1933-1938

[54] Bashforth F. In: Bashforth F, Adams JC, editors. An Attempt to Test the Theories of Capillary Action by Comparing the Theoretical and Measured Forms of Drops of Fluid, with an Explanation of the Method of Integration Employed in Constructing the Tables which Give the Theoretical Forms of Such Drops. UK: Cambridge University Press; 1883

[55] Quincke G. Ueber die Capillaritätsconstanten des Quecksilbers. Annalen der Physik. 1858; **181**:1-48

[56] Padday JF. The profiles of axially symmetric menisci. Philosophical Transactions of the Royal Society of London A: Mathematical Physical and Engineering Sciences. 1971;**269**:265-293

[57] Bernard G, Lupis C. The surface tension of liquid silver alloys: Part I. Silver-gold alloys. Metallurgical Transactions. 1971;**2**:555-559

[58] Bachmann J, Ellies A, Hartge K. Development and application of a new sessile drop contact angle method to assess soil water repellency. Journal of Hydrology. 2000;**231**:66-75

[59] Siwiec G, Oleksiak B, Smalcerz A, Wieczorek J. Surface tension of Cu-Ag alloys/Napiecie Powierzchniowe Stopów Cu-Ag. Archives of Metallurgy and Materials. 2013;**58**:193-195

[60] Tanaka T, Matsuda M, Nakao K, Katayama Y, Kaneko D, Hara S, et al. Measurement of surface tension of liquid Ga-base alloys by a sessile drop method. Zeitschrift fuer Metallkunde. 2001;**92**:1242-1246

[61] Chao KL, Lian SS, Wu CY, Chang KL. Surface tension measurement of liquid Fe-Si-B alloy by sessile drop method. Key Engineering Materials. 2014;**573**:9-12

[62] Egry I, Ricci E, Novakovic R, Ozawa S. Surface tension of liquid metals and alloys—Recent developments. Advances in Colloid and Interface Science. 2010; **159**:198-212

[63] Krasovskyy V, Naidich Y. Surface tension and specific volume of copper-titanium melts measured by the sessile drop method. Journal of Adhesion Science and Technology. 2004;**18**: 465-471

[64] Feng X, Liu L-X, Yang R-H, Zhao H-K, Liang F, Zhang C. Surface tension of molten Ni-(Cr, Co, W) alloys and segregation of elements. Transactions of Nonferrous Metals Society of China. 2008;**18**:1184-1188

[65] Guo Z, Hindler M, Yuan W, Mikula A. Density and surface tension of liquid Bi-Cu-Sn alloys. Monatshefte für Chemie-Chemical Monthly. 2012;**143**:1617-1622

[66] Guo Z, Hindler M, Yuan W, Mikula A. The density and surface tension of In-Sn and Cu-In-Sn alloys. Monatshefte für Chemie-Chemical Monthly. 2011;**142**: 579-584

[67] Giuranno D, Tuissi A, Novakovic R, Ricci E. Surface tension and density of Al-Ni alloys. Journal of Chemical & Engineering Data. 2010;**55**:3024-3028

[68] Plevachuk Y, Sklyarchuk V, Eckert S, Gerbeth G, Novakovic R. Thermophysical properties of the liquid Ga-In-Sn eutectic alloy. Journal of Chemical & Engineering Data. 2014;**59**: 757-763

[69] Guo Z, Li S, Mikula A, Yuan W. Surface tension of liquid Au-Bi-Sn alloys. Rare Metals. 2012;**31**:250-254

[70] Oleksiak B, Łabaj J, Wieczorek J, Blacha-Grzechnik A, Burdzik R. Surface

tension of Cu-Bi alloys and wettability in a liquid alloy-refractory material-gaseous phase system. Archives of Metallurgy and Materials. 2014;**59**: 282-286

[71] Kucharski M, Fima P. The surface tension and density of liquid Ag-Bi, Ag-Sn, and Bi-Sn alloys. Monatshefte für Chemie/Chemical Monthly. 2005;**136**: 1841-1846

[72] Lee JH, Lee DN. Use of thermodynamic data to calculate surface tension and viscosity of Sn-based soldering alloy systems. Journal of Electronic Materials. 2001;**30**:1112

[73] Pstruś J, Moser Z, Gąsior W. Surface properties of liquid In-Zn alloys. Applied Surface Science. 2011;**257**: 3867-3871

[74] Sklyarchuk V, Plevachuka Y, Kaban I, Novakovic R. Surface properties and wetting behavior of liquid Ag-Sb-Sn alloys. Journal of Mining and Metallurgy, Section B: Metallurgy. 2012; **48**:443-448

[75] Sharan A, Nagasaka T, Cramb A. Surface tensions of liquid Fe-Cr and Fe-Cr-N alloys. Metallurgical and Materials Transactions B. 1994;**25**:626-628

[76] Willner J, Siwiec G, Botor J. The surface tension of liquid Cu-Fe-Sb alloys. Applied Surface Science. 2010; **256**:2939-2943

[77] Fima P. Surface tension and density of liquid Sn-Ag alloys. Applied Surface Science. 2011;**257**:3265-3268

[78] Ohira C, Fujii H, Morisada Y. Temperature dependence of surface tension of Sn-Ag alloys. Journal of Materials Engineering and Performance. 2014;**23**:1561-1567

[79] Mishchuk N, Fainerman V, Kovalchuk V, Miller R, Dukhin S. Studies of concentrated surfactant solutions using the maximum bubble pressure method. Colloids and Surfaces A: Physicochemical and Engineering Aspects. 2000;**175**:207-216

[80] Drelich J, Fang C, White C. Measurement of interfacial tension in fluid-fluid systems. Encyclopedia of Surface and Colloid Science. 2002;**3**: 3158-3163

[81] Mishchuk NA, Fainerman VB, Kovalchuk VI, Miller R, Duhkin SS. Studies of concentrated surfactant solutions using the maximum bubble pressure method. Colloid Surface A. 2000;**175**:207-216

[82] Sugden S. XCVII—The determination of surface tension from the maximum pressure in bubbles. Journal of the Chemical Society, Transactions. 1922;**121**:858-866

[83] Rehbinder P. Dependence of surface activity and surface tension of solutions upon temperature and concentration. Zeitschrift für Physikalische Chemie. 1924;**111**:447-464

[84] Gasior W, Pstruś J, Moser Z, Krzyżak A, Fitzner K. Surface tension and thermodynamic properties of liquid Ag-Bi solutions. Journal of Phase Equilibria. 2003;**24**:40-49

[85] Gasior W, Moser Z, Pstruś J. Surface tension, density, and molar volume of liquid Sb-Sn alloys: Experiment versus modeling. Journal of Phase Equilibria. 2003;**24**:504-510

[86] Moser Z, Gasior W, Pstruś J. Surface tension of liquid Ag-Sn alloys: Experiment versus modeling. Journal of Phase Equilibria. 2001;**22**:254-258

[87] Liu X, Inohana Y, Takaku Y, Ohnuma I, Kainuma R, Ishida K, et al. Experimental determination and thermodynamic calculation of the phase equilibria and surface tension in the

Sn-Ag-In system. Journal of Electronic Materials. 2002;**31**:1139-1151

[88] Pstruś J, Moser Z, Gąsior W, Dębski A. Surface tension and density measurements of liquid Sn-Zn alloys. Experiment vs. SURDATdatabase of Pb-free solders. Archives of Metallurgy and Materials. 2006;**51**:335-343

[89] Moser Z, Gasior W, Bukat K, Pstrus J, Kisiel R, Sitek J, et al. Pb-free solders: Part III. Wettability testing of Sn-Ag-Cu-Bi alloys with Sb additions. Journal of Phase Equilibria and Diffusion. 2007;**28**:433-438

[90] Kraut J, Stern W. The density of gold-silver-copper alloys and its calculation from the chemical composition. Gold Bulletin. 2000;**33**: 52-55

[91] Moser Z, Gasior W, Pstrus J, Ishihara S, Liu XJ, Ohnuma I, et al. Surface tension and density measurements of Sn-Ag-Sb liquid alloys and phase diagram calculations of the Sn-Ag-Sb ternary system. Materials Transactions. 2004;**45**:652-660

[92] Gąsior W, Moser Z, Pstruś J. Measurements of the surface tension and density of TIN based Sn-Ag-Cu-Sb liquid alloys. Archives of Metallurgy and Materials. 2004;**49**:155-167

[93] Moser Z, Gasior W, Pstruś J, Ksiezarek S. Surface-tension measurements of the eutectic alloy (Ag-Sn 96.2 at.%) with Cu additions. Journal of Electronic Materials. 2002;**31**: 1225-1229

[94] Moser Z, Gasior W, Pstrus J. Surface tension measurements of the Bi-Sn and Sn-Bi-Ag liquid alloys. Journal of Electronic Materials. 2001;**30**:1104-1111

[95] Moser Z, Gąsior W, Bukat K, Pstruś J, Kisiel R, Sitek J, et al. Pb-free solders: Part 1. Wettability testing of Sn-Ag-Cu alloys with Bi additions. Journal of Phase Equilibria and Diffusion. 2006;**27**: 133-139

[96] Moser Z, Gasior W, Ishida K, Ohnuma I, Bukat K, Pstrus J, et al. Experimental wettability studies combined with the related properties from data bases for lead-free soldering materials. In: CALPHAD XXXIII Program & Abstract; 2004

[97] Moser Z, Gąsior W, Pstruś J, Ohnuma I, Ishida K. Influence of Sb additions on surface tension and density of Sn-Sb, Sn-Ag-Sb and Sn-Ag-Cu-Sb alloys: Experiment vs. modeling. Zeitschrift fuer MetaIlkunde. 2006;**97**: 365-370

[98] Gasior W, Moser Z, Pstruś J, Bukat K, Sitek J, Kisiel R. (Sn-Ag)$_{eut}$+Cu soldering materials, part I: Wettability studies. Journal of Phase Equilibria and Diffusion. 2004;**25**:115-121

[99] Moser Z, Gasior W, Pstrus J, Ksiezarek S. Surface tension and density of the (Ag-Sn) eut+Cu liquid alloys. Journal of Electronic Materials. 2002;**31**: 1225-1229

[100] Moser Z, Gąsior W, Ishida K, Ohnuma I, Liu X, Bukat K, et al. Experimental wettability studies combined with the related properties from data base for tin based alloys with silver, copper, bismuth and antimony additions. In: 134th Annual Meeting & Exhibition, Book of Final Program; 2005. pp. 13-17

[101] Moser Z, Gasior W, Debski A. Database of Pb-free soldering materials, surface tension and density, experiment vs. modeling. Data Science Journal. 2005;**4**:195-208

[102] Gasior W, Moser Z, Bukat K, Pstrus J, Sitek J, Kisiel R. Experimental wettability studies of tin based liquid alloys with Ag, Cu, Bi and Sb additions. TOFA; 2004

[103] Jacob KT, Jayadevan KP. (Sn-Ag)$_{eut}$+Cu soldering materials, part II: Electrical and mechanical studies. Journal of Phase Equilibria and Diffusion. 2004;**25**:122-124

[104] Gasior W, Pstrus J, Moser Z. Influence of zinc on the surface tension, density and molar volume of (Ag-Sn) (EUT)+ Zn liquid alloys. Archives of Metallurgy and Materials. 2016;**61**: 361-367

[105] Goicoechea J, Garcia-Cordovilla C, Louis E, Pamies A. Surface tension of binary and ternary aluminium alloys of the systems Al-Si-Mg and Al-Zn-Mg. Journal of Materials Science. 1992;**27**: 5247-5252

[106] Pstruś J. Surface tension and density of liquid In-Sn-Zn alloys. Applied Surface Science. 2013;**265**:50-59

[107] Quested P, Brooks R, Chapman L, Morrell R, Youssef Y, Mills K. Measurement and estimation of thermophysical properties of nickel based superalloys. Materials Science and Technology. 2009;**25**:154-162

[108] Mukai K, Matsushita T, Mills KC, Seetharaman S, Furuzono T. Surface tension of liquid alloys—A thermodynamic approach. Metallurgical and Materials Transactions B. 2008;**39**: 561-569

[109] Mills K, Su Y. Review of surface tension data for metallic elements and alloys: Part 1-pure metals. International Materials Reviews. 2006;**51**:329-351

[110] Mills KC, Youssef YM, Li Z, Su Y. Calculation of thermophysical properties of Ni-based superalloys. ISIJ International. 2006;**46**:623-632

[111] Aune R, Battezzati L, Brooks R, Egry I, Fecht H-J, Garandet J-P, et al. Surface tension and viscosity of industrial alloys from parabolic flight experiments—Results of the thermo lab project. Microgravity Science and Technology. 2005;**16**:11-14

[112] Aune R, Battezzati L, Brooks R, Egry I, Fecht H-J, Garandet J-P, et al. Measurement of thermophysical properties of liquid metallic alloys in a ground-and microgravity based research programme—The ThermoLab project. Microgravity Science and Technology. 2005;**16**:7-10

[113] Li Z, Mukai K, Zeze M, Mills K. Determination of the surface tension of liquid stainless steel. Journal of Materials Science. 2005;**40**:2191-2195

[114] Su Y, Li Z, Mills K. Equation to estimate the surface tensions of stainless steels. Journal of Materials Science. 2005;**40**:2201-2205

[115] Su Y, Mills K, Dinsdale A. A model to calculate surface tension of commercial alloys. Journal of Materials Science. 2005;**40**:2185-2190

[116] Mills KC, Su Y, Li Z, Brooks RF. Equations for the calculation of the thermo-physical properties of stainless steel. ISIJ International. 2004;**44**: 1661-1668

[117] Li Z, Mills KC, McLean M, Mukai K. Measurement of the density and surface tension of Ni-based superalloys in the liquid and mushy states. Metallurgical and Materials Transactions B. 2005;**36**:247-254

[118] Mills K, Brooks R. Measurements of thermophysical properties in high temperature melts. Materials Science and Engineering A. 1994;**178**:77-81

[119] Margrave JL. Determination of thermophysical properties of liquid metals at high temperatures by levitation methods. Materials Science and Engineering A. 1994;**178**:83-88

[120] Brooks R, Monaghan B, Barnicoat A, McCabe A, Mills K, Quested P. The

physical properties of alloys in the liquid and "mushy" states. International Journal of Thermophysics. 1996;**17**: 1151-1161

[121] Brooks R, Day A, Mills K, Quested P. Physical property measurements for the mathematical modeling of fluid flow in solidification processes. International Journal of Thermophysics. 1997;**18**: 471-480

[122] Keene B, Mills K, Kasama A, McLean A, Miller W. Comparison of surface tension measurements using the levitated droplet method. Metallurgical Transactions B. 1986;**17**:159-162

[123] Keene B, Mills K, Brooks R. Surface properties of liquid metals and their effects on weldability. Materials Science and Technology. 1985;**1**:559-567

[124] Keene BJ, Mills K, Bryant J, Hondros E. Effects of interaction between surface active elements on the surface tension of iron. Canadian Metallurgical Quarterly. 1982;**21**: 393-403

[125] Zeng BFG. Influencing the wettability of HSS-steel by addition of alloyingelmentsto the zink bath. In: Proceeding of EMC; 2009. pp. 1-18

[126] Adhikari D, Jha I, Singh B. Transport and surface properties of molten Al-Mn alloy. Advanced Materials Letters. 2012;**3**:226-230

[127] Brillo J, Lauletta G, Vaianella L, Arato E, Giuranno D, Novakovic R, et al. Surface tension of liquid Ag-Cu binary alloys. ISIJ International. 2014; **54**:2115-2119

[128] Mills KC. Recommended Values of Thermophysical Properties for Selected Commercial Alloys. UK: Cambridge University Press; 2002

[129] Ohnuma I, Ishida K, Moser Z, Gsior S, Bukat K, Pstru J, et al. Pb-free solders: Part II. Application of ADAMIS database in modeling of Sn-Ag-Cu alloys with Bi additions. Journal of Phase Equilibria and Diffusion. 2006;**27**:245

[130] Liu XJ, Yamaki T, Ohnuma I, Kainuma R, Ishida K. Thermodynamic calculations of phase equilibria, surface tension and viscosity in the In-Ag-X (X = Bi, Sb) system. Materials Transactions. 2004;**45**:637-645

[131] Hoffmann M, Wynblatt P. Surface composition of ternary Cu-Ag-Au alloys: Part I. Experimental results. Metallurgical Transactions A. 1991;**22**: 1833-1840

[132] Gancarz T. Physicochemical properties of Sb-Sn-Zn alloys. Journal of Electronic Materials. 2014;**43**:4374-4385

[133] Yadav S, Jha L, Jha I, Singh B, Koirala R, Adhikari D. Prediction of thermodynamic and surface properties of Pb-Hg liquid alloys at different temperatures. Philosophical Magazine. 2016;**96**:1909-1925

[134] Gancarz T, Pstru J, Gsior W, Henein H. Physicochemical properties of Sn-Zn and SAC+ Bi alloys. Journal of Electronic Materials. 2013;**42**:288

[135] Egry I. The surface tension of binary alloys: Simple models for complex phenomena. International Journal of Thermophysics. 2005;**26**: 931-939

[136] McNallan MJ, Debroy T. Effect of temperature and composition on surface tension in Fe-Ni-Cr alloys containing sulfur. Metallurgical Transactions B. 1991;**22**:557-560

[137] Novakovic R, Tanaka T. Bulk and surface properties of Al-Co and Co-Ni liquid alloys. Physica B: Condensed Matter. 2006;**371**:223-231

[138] Costa C, Delsante S, Borzone G, Zivkovic D, Novakovic R.

Thermodynamic and surface properties of liquid Co-Cr-Ni alloys. The Journal of Chemical Thermodynamics. 2014;**69**: 73-84

[139] Gąsior W, Fima P, Moser Z. Modeling of the thermodynamic properties of liquid Fe-Ni and Fe-Co alloys from the surface tension data. Archives of Metallurgy and Materials. 2011;**56**:13-23

[140] Yan L, Zheng S, Ding G, Xu G, Qiao Z. Surface tension calculation of the Sn-Ga-In ternary alloy. Calphad. 2007;**31**:112-119

[141] Trybula M, Gancarz T, Gąsior W. Density, surface tension and viscosity of liquid binary Al-Zn and ternary Al-Li-Zn alloys. Fluid Phase Equilibria. 2016; **421**:39-48

[142] Tanaka T. Prediction of phase diagrams in nano-sized binary alloys. Materials Science Forum. 2010;**653**:55-75

[143] Prasad L, Mikula A. Surface segregation and surface tension in Al-Sn-Zn liquid alloys. Physica B: Condensed Matter. 2006;**373**:142-149

[144] Novakovic R, Brillo J. Thermodynamics, thermophysical and structural properties of liquid Fe-Cr alloys. Journal of Molecular Liquids. 2014;**200**:153-159

[145] Prasad L, Xie Y, Mikula A. Lead free solder materials In-Sn-Zn system. Journal of Non-Crystalline Solids. 1999; **250**:316-320

[146] Terzieff P, Li Z, Knott S, Mikula A. Physicochemical properties of liquid Ag-Bi-Sn. Physica B: Condensed Matter. 2007;**388**:312-317

[147] Terzieff P. Surface tension, viscosity and concentration fluctuations in liquid Ag-In-Sn. Physics and Chemistry of Liquids. 2006;**44**:115-125

[148] Terzieff P. Concentration fluctuations and surface tension in liquid Au-Sn-Zn. Journal of Materials Science. 2005;**40**:3759-3763

[149] Wang X, Li W. Models to estimate viscosities of ternary metallic melts and their comparisons. Science in China, Series B: Chemistry. 2003;**46**:280-289

[150] Liu X-j, Wang C-p, Ohnuma I, Kainuma R, Ishida K. Development of thermodynamic and kinetic databases in micro-soldering alloy systems and their applications. Progress in Natural Science: Materials International. 2011;**21**:97-110

[151] Prasad L, Jha R. Correlation between bulk and surface properties of ternary Ag-Sn-Zn liquid alloys and concerned binaries. Physics and Chemistry of Liquids. 2007;**45**:149-167

[152] Kaban I, Gruner S, Hoyer W. Surface tension and density in liquid Ag-Cu-Sn alloys. Journal of Non-Crystalline Solids. 2007;**353**:3717-3721

[153] Kaban IG, Gruner S, Hoyer W. Experimental and theoretical study of the surface tension in liquid Ag-Cu-Sn alloys. Monatshefte für Chemie/Chemical Monthly. 2005;**136**:1823-1828

[154] Akinlade O, Singh R. Bulk and surface properties of liquid In-Cu alloys. Journal of Alloys and Compounds. 2002; **333**:84-90

[155] Ohnuma I, Liu X, Ohtani H, Ishida K. Thermodynamic database for phase diagrams in micro-soldering alloys. Journal of Electronic Materials. 1999;**28**: 1164-1171

[156] Tanaka T, Hack K, Hara S. Use of thermodynamic data to determine surface tension and viscosity of metallic alloys. MRS Bulletin. 1999;**24**:45-51

[157] Wang L, Chou K-C, Seetharaman S. A comparison of traditional geometrical models and mass triangle model in

calculating the surface tensions of ternary sulphide melts. Calphad. 2008;**32**:49-55

[158] Kaptay G. A method to calculate equilibrium surface phase transition lines in monotectic systems. Calphad. 2005;**29**:56-67

[159] Novakovic R, Giuranno D, Delsante S, Borzone G. Bulk and surface properties of liquid Cr-Nb-Re alloys. Journal of Phase Equilibria and Diffusion. 2014;**35**:445-457

[160] Taskinen J, Yliruusi J. Prediction of physicochemical properties based on neural network modelling. Advanced Drug Delivery Reviews. 2003;**55**: 1163-1183

[161] Qiao Z, Yan L, Cao Z, Xie Y. Surface tension prediction of high-temperature melts. Journal of Alloys and Compounds. 2001;**325**:180-189

[162] Yeum K, Speiser R, Poirier DR. Estimation of the surface tensions of binary liquid alloys. Metallurgical Transactions B. 1989;**20**:693-703

[163] Picha R, Vřešťál J, Kroupa A. Prediction of alloy surface tension using a thermodynamic database. Calphad. 2004;**28**:141-146

[164] Bradshaw R, Warren M, Rogers J, Rathz T, Gangopadhyay A, Kelton K, et al. Containerless measurements of thermophysical properties of $Zr_{54}Ti_8Cu_{20}Al_{10}Ni_8$. Annals of the New York Academy of Sciences. 2006;**1077**: 63-74

[165] Hoffmann M, Wynblatt P. Surface composition of ternary Cu-Ag-Au alloys: Part II. A comparison of experiment with theoretical models. Metallurgical Transactions A. 1991;**22**: 1841-1848

[166] Gallois B, Lupis C. Surface tensions of liquid Ag-Au-Cu alloys. Metallurgical Transactions B. 1981;**12**:679-689

[167] Pajarre R, Koukkari P, Tanaka T, Lee J. Computing surface tensions of binary and ternary alloy systems with the Gibbsian method. Calphad. 2006;**30**: 196-200

[168] Hu J, Chen S, Cai W, Chou K, Zhou G. Predicting the surface tension of ternary by using geometric solution model. Journal of University of Science and Technology Beijing. 1990;**12**: 558-562

[169] Calvo F. Molecular dynamics determination of the surface tension of silver-gold liquid alloys and the Tolman length of nanoalloys. The Journal of Chemical Physics. 2012;**136**:154701

[170] Benisek A, Dachs E. The vibrational and configurational entropy of disordering in Cu_3Au. Journal of Alloys and Compounds. 2015;**632**: 585-590

[171] Muggianu YM, Gambino M, Bros J. Enthalpies of formation of liquid alloys bismuth-gallium-tin at 723k-choice of an analytical representation of integral and partial thermodynamic functions of mixing for this ternary-system. Journal de Chimie Physique et de Physico-Chimie Biologique. 1975;**72**:83-88

[172] Kohler F. Zur Berechnung der thermodynamischen Daten eines ternären systems aus den zugehörigen binären Systemen. Monatshefte für Chemie und verwandte Teile anderer Wissenschaften. 1960;**91**:738-740

[173] Toop G. Predicting ternary activities using binary data. Transactions of the Metallurgical Society of AIME. 1965;**233**:850

[174] Chou K-C. A general solution model for predicting ternary thermodynamic properties. Calphad. 1995;**19**:315-325

[175] Butler J. The thermodynamics of the surfaces of solutions. Proceedings of

the Royal Society of London. Series A, Containing Papers of a Mathematical and Physical Character. 1932;**135**: 348-375

[176] Dogan A, Arslan H, Dogan T. Estimation of excess energies and activity coefficients for the penternary Ni-Cr-Co-Al-Mo system and its subsystems. The Physics of Metals and Metallography. 2015;**116**:544-551

[177] Dogan A, Arslan H. Comparative thermodynamic prediction of integral properties of six component, quaternary, and ternary systems. Metallurgical and Materials Transactions A. 2015;**46**:3753-3760

[178] Arslan H. Determinations of enthalpy and partial molar enthalpy in the alloys Bi-Cd-Ga-In-Zn, Bi-Cd-Ga-Zn and Au-Cu-Sn. Materials Chemistry and Physics. 2015;**153**:384-389

[179] Arslan H, Dogan A. An analytical investigation for thermodynamic properties of the Fe-Cr-Ni-Mg-O system. Russian Journal of Physical Chemistry A. 2015;**89**:180-189

[180] Arslan H, Dogan A, Dogan T. An analytical approach for thermodynamic properties of the six-component systems Ni-Cr-Co-Al-Mo-Ti and their subsystems. The Physics of Metals and Metallography. 2013;**114**:1053-1060

[181] Arslan H. Analytical determination of partial and integral properties of the six components systems Ni-Cr-Co-Al-Mo-Ti and their subsystems. Physica B: Condensed Matter. 2014;**438**:48-52

[182] Dogan A, Arslan H. An investigation on surface tensions of Pb-free solder materials. Philosophical Magazine. 2016;**96**:2887-2901

[183] Arslan H, Dogan A. Thermodynamic investigations on the component dependences of high-entropy alloys. Russian Journal of Physical Chemistry A. 2016;**90**: 2339-2345

[184] Tanaka T, Hack K, Iida T, Hara S. Application of thermodynamic databases to the evaluation of surface tensions of molten alloys, salt mixtures and oxide mixtures. Zeitschrift fuer Metallkunde. 1996;**87**:380-389

[185] Iida T, Guthrie RI. The Physical Properties of Liquid Metals. Oxford, UK: Clarendon Press; 1988

[186] Moser Z, Gasior W, Pstrus J, Zakulski W, Ohnuma I, Liu XJ, et al. Studies of the Ag-In phase diagram and surface tension measurements. Journal of Electronic Materials. 2001;**30**: 1120-1128

[187] Moser Z, Gąsior W, Pstruś J, Dębski A. Wettability studies of Pb-free soldering materials. International Journal of Thermophysics. 2008;**29**: 1974-1986

[188] Kuntz M, Panton B, Wasiur-Rahman S, Zhou Y, Corbin S. An experimental study of transient liquid phase bonding of the ternary Ag-Au-Cu system using differential scanning calorimetry. Metallurgical and Materials Transactions A. 2013;**44**:3708-3720

[189] Prasad L, Mikula A. Role of surface properties on the wettability of Sn-Pb-Bi solder alloys. Journal of Alloys and Compounds. 1999;**282**:279-285

[190] Guggenheim EA. Mixtures: The Theory of the Equilibrium Properties of some Simple Classes of Mixtures Solutions and Alloys. Oxford, UK: Clarendon Press; 1952

[191] Fowler RH, Guggenheim EA. Statistical Thermodynamics: A Version of Statistical Mechanics for Students of Physics and Chemistry. Cambridge: Cambridge University Press; 1960

[192] Singh R. Short-range order and concentration fluctuations in binary

molten alloys. Canadian Journal of Physics. 1987;**65**:309-325

[193] Prasad L, Singh R, Singh V, Singh G. Correlation between bulk and surface properties of AgSn liquid alloys. The Journal of Physical Chemistry B. 1998; **102**:921-926

[194] Bhatia A, Singh R. Short range order and concentration fluctuations in regular and compound forming molten alloys. Physics and Chemistry of Liquids an International Journal. 1982;**11**:285-313

[195] Bhatia A, Singh R. Thermodynamic properties of compound forming molten alloys in a weak interaction approximation. Physics and Chemistry of Liquids an International Journal. 1982;**11**:343-351

[196] Novakovic R, Tanaka T, Muolo M, Lee J, Passerone A. Bulk and surface properties of liquid Ag-X (X = Ti, Hf) compound forming alloys. Surface Science. 2005;**591**:56-69

[197] Kaptay G, Papp Z. On the concentration dependence of the surface tension of liquid metallic alloys theoretical basis. In: Proceedings of Microcad, Materials Science and Technology, University of Miskolc; 2002. pp. 45-50

[198] Hassam S, Ägren J, Gaune-Escard M, Bros J. The Ag-Au-Si system: Experimental and calculated phase diagram. Metallurgical Transactions A. 1990;**21**:1877-1884

[199] Kusoffsky A. Thermodynamic evaluation of the ternary Ag-Au-Cu system—Including a short range order description. Acta Materialia. 2002;**50**: 5139-5145

[200] Sundman B, Fries SG, Oates WA. A thermodynamic assessment of the Au-Cu system. Calphad. 1998;**22**: 335-354

[201] Yan L, Cao Z, Xie Y, Qiao Z. Surface tension calculation of the Ni_3S_2 FeS Cu_2S mattes. Calphad. 2000;**24**: 449-463

[202] Qiao Z-Y, Xing X, Peng M, Mikula A. Thermodynamic criterion for judging the symmetry of ternary systems and criterion applications. Journal of Phase Equilibria. 1996;**17**:502-507

[203] Available from: https://commons.wikimedia.org/wiki/File:Ag-Au-Cu-colours-english.svg

Section 4

Mechanical Property of Composites

Chapter 6

Microstructure and Property Co-relation of 4% Carbon Fe-C Alloy System

Ranjan Kumar Behera, Sudipta Sen and Subash Chandra Mishra

Abstract

The suitable combination of strength and toughness makes ductile cast iron (DCI) the most preferred alloy for structural and engineering applications, among other members in the Fe-C alloy system. The property of DCI is dependent upon its microstructure that varies with dosage of alloying elements during the casting process as well as upon subjected to different heat treatment process. The microstructure of DCI consists of spheroidal graphite and either a single phase or combination of phases depending upon alloying element and heat treatment process, consequentially resulting in a combination of strength and toughness as desired for specified application. The current chapter discusses the relation between change in microstructural aspects with respect to various alloying elements as well as heat treatment processes. Emphasis is given on reveling the effect of physical characteristics with mechanical and tribological properties. The chapter is concluded by a case study on the effect of alloying elements and heat treatment processes on microstructure as well as mechanical and tribological properties of DCI.

Keywords: ductile cast iron, alloying, microconstituent, toughness, tribology

1. Introduction

The iron-carbon (Fe-C) system is a binary alloy system that comprises of 6.67% (max.) of carbon in iron, which is further classified into two alloys i.e., steel comprises of maximum 2% of C and cast iron having carbon that ranges from 2 to 6.67%. The cast iron is further classified into white cast iron and gray cast iron depending on the microstructure so obtained after the casting process. However the desired morphology to be obtained at room temperature can only be modified at the time of inoculation during the melting process by adding appropriate amount of alloying elements. Steel is the most preferable candidate for structural application as it possesses a good combination of strength, hardness, ductility, and impact toughness, whereas cast iron finds its application where the system is subjected to vibration such as base for heavy machinery installation. In structural application cast iron does not sit suitably due to its brittle nature that allows rapid fracture growth, having a high strength but low ductility and low impact toughness. The reason for such behavior is the nature of graphite in the final microstructure which remains in the form of flakes with sharp edges that result in stress concentration sites upon loading and ultimately quicker crack propagation.

Spheroidal graphite cast iron (SGCI)/ductile cast iron (DCI) as the name suggests, has graphite in spherical form embedded in the matrix. The spherical shape is achieved by the addition of magnesium (Mg) and/or cerium (Ce) as inoculants, to the gray cast iron melt during the casting process. The graphite spheroids do not interfere with matrix continuity and eliminate the stress concentration effect that generates due to the presence of flaky graphite in case of gray cast iron, leading to improved ductility and impact toughness without affecting the strength and hardness. In addition, to these advantages, it possesses ease of castability, machinability and less cost of production. Because of these enhanced properties over the other cast irons, during the last five decades, the production and commercialization of SGCI have grown immensely. The versatility of SGCI in strength, toughness, tribological behavior and resistance to corrosion made it the preferable candidate for structural, automotive, agricultural and many more application including nuclear industry as well. Though in as-received form via melting and casting, it discloses desired property combination for the particular application, the properties can further be improved by employing suitable heat treatment techniques. The austempering heat treated ductile iron (refereed as ADI) shows excellent combination of high strength and toughness [1, 2] along with better wear resistance [3, 4], excellent fatigue strength [5, 6] as well as fracture toughness [7, 8]. Application of suitable heat treatment process leads to transformation of as-cast ferritic or ferritic/pearlitic matrix into fully pearlitic, martensitic, and bainitic resulting variation in the properties. Among all the heat treatment processes employed the austempering treatment resulting in bainitic matrix is the most favorable member due to its strength, impact toughness, tribological behavior as well as fatigue and fracture toughness [9–13]. Tribological properties of austempered ductile iron (ADI) is observed to be in direct relationship with the austempering temperature and time. The consequences being increased hardness during the wear attributed to the fact that, the bainitic ferrite in the matrix undergoes strain hardening as it is less prone to thermal instability as compared to martensite [14]. On the other hand the tempering treatment with Boron alloyed DI reported to result in increased wear resistance; however, it decreases with increased boriding time [15]. Not only the matrix microstructure and alloying element is responsible for the wear behavior, but also as reported by Sugishita and Fujiyoshi [16]. Zimba et al. [17] the graphite characteristics viz. nodule size and distribution in the matrix has significant effect on the wear properties as well.

2. Effect of alloying element on morphological and mechanical properties of SGCI

To achieve desired properties for a particular application, control over matrix structure and morphological aspects is necessary, and that can be achieved by controlling processing parameters and the addition of alloying elements. The addition of alloying elements influences the mechanical properties by changing the cast matrix microstructure from ferritic to pearlitic, ferritic/pearlitic and/or austenitic depending upon the alloy composition. If the processing parameters such as tapping temperature and cooling rate are kept constant, the final matrix bends toward the matrix influencing element. Incorporation of Ni from 12 to 38% (max.), a highly ductile material of austenitic SGCI is obtained. On the other hand, a higher Si amount results in ferritic matrix and that of Cu and Mn lead to the formation of harder pearlitic matrix. A proper control of ferrite and pearlite influencing elements results in a mixed ferritic/pearlitic structure having properties intermediate between ferritic and pearlitic grades of SGCI.

2.1 Effect of base composition

Silicon is the most vital element in the production of SGCI and influences the solidification process in conjunction with Carbon. An increase of 1% Si content shifts the eutectic composition toward the left (approximately 0.30% of carbon), resulting in low solidification temperature. With increasing Si content the eutectoid carbon content is lowered, and hence transformation occurs over a broadening range. **Figure 1(a)** depicts the inverse relationship of carbide content in the matrix with silicon content, and as disclosed with increased Si content there is a decrease in the carbide formation. Silicon is a graphitizer and ferrite promoter, hence reduces the strength properties of ferritic ductile iron, which if needed, further can be improved by the addition of desired amount of nickel. Increased silicon content increases the carbon equivalent (%CE), hence prevents the precipitation of carbides and allows the formation of more free form of carbon leading to increased nodule count (**Figure 1(b)**) and decreased hardness and UTS [18–20]. On the other hand, for austenitic grades increase in Si content increases the tensile strength and 0.2% yield strength [18]. The substantial influence of silicon on the ductile-brittle transition temperature of ferritic ductile iron is shown in **Figure 2** [21]. From **Figure 1**, it is clearly understood that to achieve optimum low-temperature toughness, the amount of Si should be maintained as low as possible. Fully ferritic carbide free ductile iron production needs high purity charge materials, proper holding and casting practice and highly effective inoculation for maximum nodule count. Higher amount of silicon leads to a reduction in low impact toughness, increased DBTT and decreased thermal conductivity. Si is held below 2.2% when producing the ferritic grades and between 2.5 and 2.8% when producing pearlitic grades [19, 20].

Manganese is a very strong carbide stabilizer at eutectoid graphitization and increases the amount of combined carbon. Excess Mn has little effect on solidification and only weakly retards primary graphitization. The presence of a correct amount of Mn forms manganese sulfide and reduces the proportion of combined carbon by removing the effect of sulfur [23]. Being a mild perlite promoter it results in increased hardness and proof stress but to a smaller extent. Manganese is added to the liquid melt to a limited amount as because it prevents the initiation of eutectoid transformation leading to lower rate of carbon diffusion in the ferrite and stabilizing cementite (Fe_3C). However, the noticeable problem with Mn addition is that

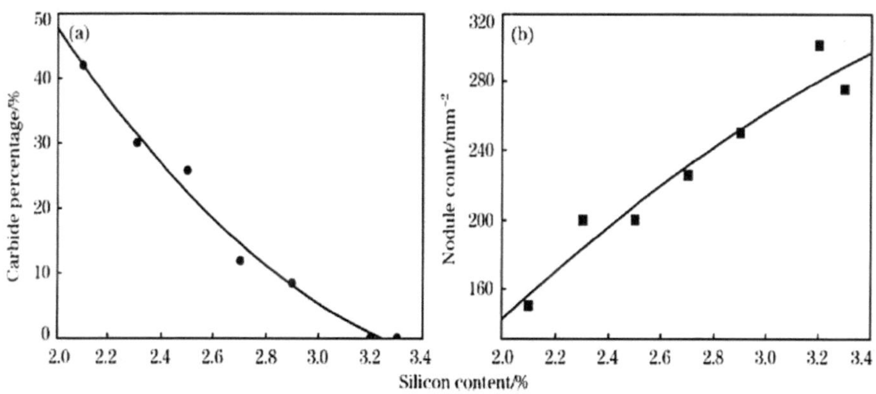

Figure 1.
Variation of carbide percentage (a) and nodule count (b) in ductile cast iron with silicon.

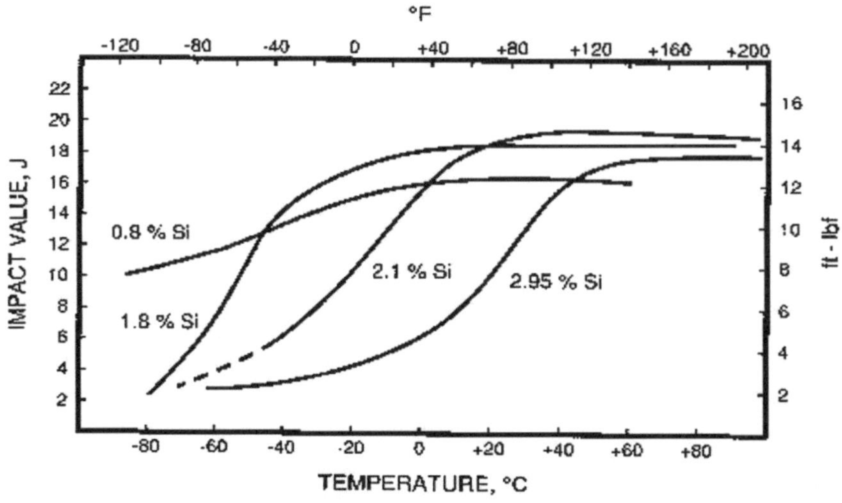

Figure 2.
Influence of silicon content on the v-notched Charpy energy of ferritic ductile iron [22].

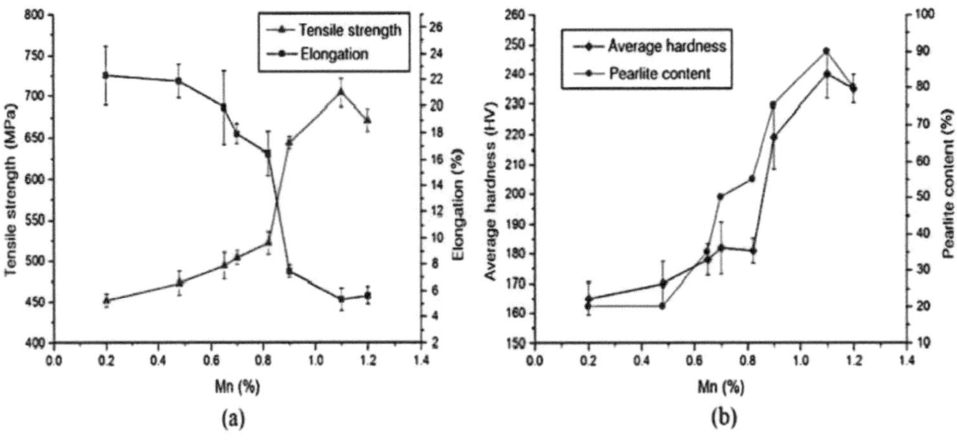

Figure 3.
(a) The relationship between tensile strength and elongation versus Mn content (0.05% Cu), (b) average hardness and pearlite content versus Mn content (0.05% Cu) [27].

the embrittlement caused during the solidification process and hence limiting its range to (0.3–1.01) [24–26]. Mn is held between 0.4 and 0.6% when making pearlitic grades and below 0.3% when making ferritic grades and can be achieved by dilution with high purity pig iron to avoid pearlite and carbide formation [25, 26]. Higher Mn content leads to increased pearlite content resulting in increased strength and hardness and decreased elongation, **Figure 3**. However for a ferritic grade of SGCI Mn dissolved into the ferrite matrix and improves strength [27]. Use of higher Mn level is not preferred as it causes segregation at the grain boundaries that causes grain boundary carburet and creation of secondary phase along the boundaries leading to degradation in plasticity in SG cast iron. In combination with Ni, higher manganese content can stabilize austenite and increases carbide by depressing the solidification cooling curve encouraging graphite to segregate at a lower temperature consequentially resulting lower generation of free graphite and reduced ductility and impact toughness [28–31].

Sulfur affects the form of carbon in a manner which is opposite to that of Si. The higher the S content, the higher will be the amount of combined carbon, thus tends to form a hard and brittle matrix. An excessive amount of S will increase the tendency of dross defects above 0.015% promote the formation of quasi-flaky graphite [23, 26]. The addition of S is done for better machinability, but it is kept around 0.009, and maximum 0.015% as the larger additions of Sulfur may cause the hot (red) shortness [24]. Sulfur is an active interfacial element and has less solid solubility in austenite, therefore, gets enriched in liquid melt and at the freezing point. The presence of S encourages the eutectic graphite to branch frequently resulting in formation of chunky graphite [32].

Phosphorus combines with iron to form iron phosphide and produces a ternary eutectic with cementite and austenite. The ternary eutectic is known as steadite, which is relatively brittle and tends to form a continuous network outlining the primary austenite dendrites resulting lower toughness. It increases fluidity and extends the range of eutectic freezing, thus increasing primary graphitization when Si content is high. Incorporation of P above 0.05% can produce internal defects. It also causes embrittlement of iron, increases the ductile-brittle transition temperature, promotes galvanizing and temper-embrittlement. P often segregates to grain boundaries and produces carbide/phosphide complexes, hence needs to be maintained as low as possible. P is kept intentionally low as it causes cold shortness and so the property of ductile iron will be ruined [24].

Magnesium is used as nodulizer and responsible for spherical form of graphite, but at lower level chunky graphites are formed. Like sulfur, it is also an interfacial element and has less solid solubility in austenite. Magnesium makes the eutectic graphite grow with screw dislocations along the crystal orientation, giving the eutectic graphite growth characteristic into a spherical form. The lower level of Mg counteracts the surface activity of sulfur and oxygen promoting the formation and growth of chunky graphite particles. Due to the segregation effect when Magnesium content becomes high enough the graphite shape converts to spherical form [32].

2.2 Effect of alloying element

Chromium prevents the corrosion by forming the layer of chromium oxide on the surface and stops the further exposition of the surface to the atmosphere. However, as it is a strong carbide former, it is not required in carbide free structure. In general it is kept around 0.05%. It must be kept <1% in GGG-50 grade [24].

Nickel is used for strengthening ferrite in addition to Si leading to increased UTS without compensating ductility and impact values. It is usually added in traces due to high cost and also to avoid the embrittlement of matrix (if it exceeds 2%) [24]. Increased Ni content decreases the ductility for austenitic grade SGCI [33]. Nickel assists in shifting the transformation temperature boundary leading to attention of desired phase at lower temperature when alloyed with Ni as compared to that for an alloy without Ni. Additionally it has the ability to prevent the precipitation of secondary carbides in the final microstructure. Uma et al. [34] studied the effect of toughened austempered SG cast iron with increasing Ni content up to 2.5%. It is reported in their work that, with increasing Ni content pearlite content is increased in the as-cast matrix. Also, the retained austenite gets stabilized in the final microstructure. It is increased due to the fact that Ni concentrates in the austenite phase in the $(\alpha + \gamma)$ region, resulting in increased impact toughness and wear resistance. When subjected to austempering treatment SG cast iron with Ni content higher than 0.5%, slows down the bainitic reaction and causes the formation of martensite at the austenite cell boundaries on cooling [32]. Although the addition

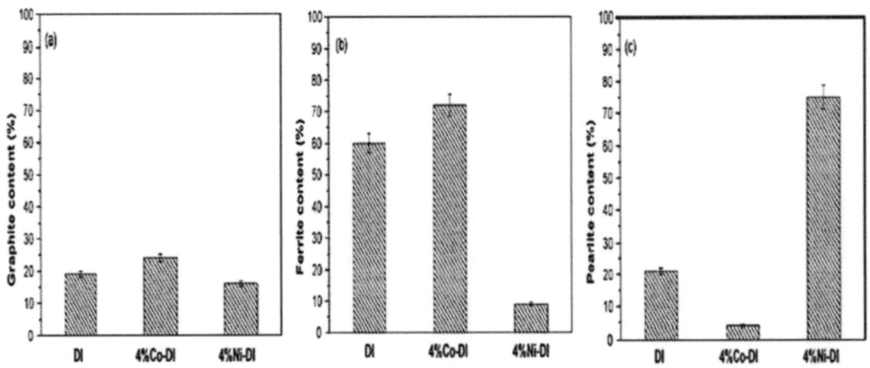

Figure 4.
Comparison of constituent content in the microstructure of the Co and Ni alloyed SG cast irons: (a) graphite, (b) ferrite and (c) pearlite [35].

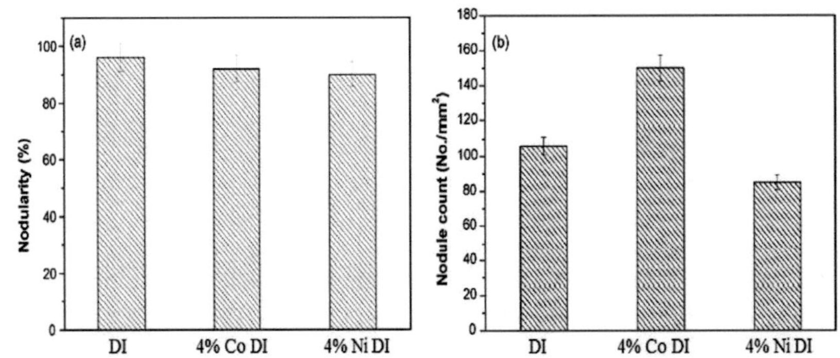

Figure 5.
Comparison of nodular graphite of the Co and Ni alloyed SG cast irons: (a) nodularity and (b) nodule count [35].

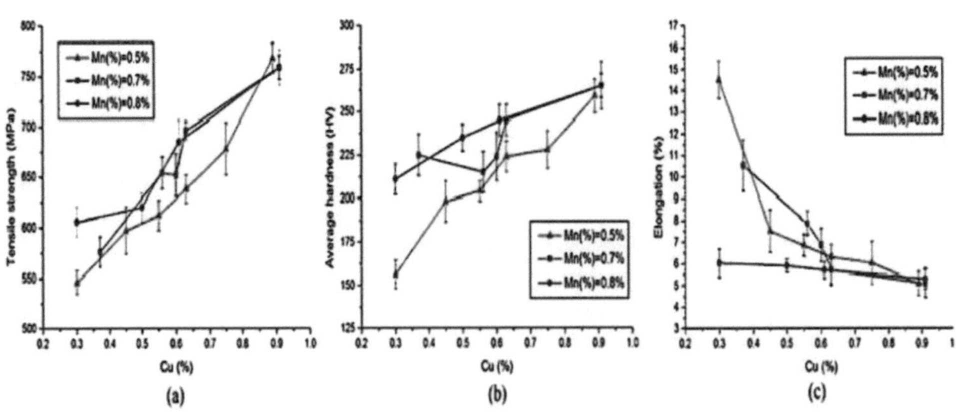

Figure 6.
Relationship between the mechanical properties and the copper content for different Mn level [27]. (a) Mn content vs. Tensile Strength, (b) Mn content vs. Average Hardness, (c) Mn content vs. % Elongation.

of Ni promotes the stabilization of austenitic matrix in SG cast iron, it accelerates the formation of pearlite by shifting the knee of transformation diagram to higher time intervals. Thus, the eutectoid transformation of ductile cast iron in cooling austenite has resulted in a significant amount of pearlite structure (**Figure 4**) and decreasing nodularity and nodule count (**Figure 5**) with increased yield and tensile strength and hardness with decreased ductility as well as impact toughness [35].

Copper is a strong pearlite promoter and is undesirable in ferritic grade. It increases the UTS, 0.2% yield strength and hardness (**Figure 6**) without an embrittlement in the matrix. In total Cu is kept between 0.4–0.8% [24, 25]. Along with promoting, it also refines the slice distance of pearlite in the eutectoid transformation period and is beneficial in stabilizing super-cooled austenite and increases the closing rate of the austenite shell that enables graphite to achieve perfectly spherical shape [27, 32, 36]. Copper strengthens the matrix when Mn levels are low. Unlike Mn, copper has an adverse effect on segregation and usually gathers at the interface between graphite and matrix. As a result, the diffusion of C is restricted due to a higher concentration of Cu, which leads to the formation of perfectly spherical nodules and increased nodule count. The combined effect of Cu and Ni, when subjected to austempering treatment, balances the hardenability, segregation tendency of Mn and inhibits the transformation of untransformed austenite into pearlite during quenching to austempering temperature. An excess amount of Cu results in the evolution of copper or copper-rich phase that affects the mechanical properties [37].

Molybdenum is a mild pearlite promoter and forms intercellular carbides especially in heavy sections leading to increased proof stress and hardness and also improves properties at elevated temperature [24]. Also, Mo segregates at the grain boundaries causing decrease in impact toughness. When subjected to austempering or ausforming process increased amount of Molybdenum stabilizes and increases austenite and bainite content. At zero Mo content the matrix is mainly bainitic, but with the increased amount it also increases the amount of carbide [18]. Mo, with an extreme tendency of segregating to the last solidification area, retards the bainite reaction and causes the micro shrinkage porosity in the intercellular region. Consequently, the Mo alloyed iron offers the lowest impact strength [21].

Rare earth (*RE*) materials in the nodulizer can eliminate the bad effect of the interferential elements like titanium, bismuth, arsenic, antimony, etc. and reduce the content of Si and Mg, which is helpful to avoid graphite degeneration. Ferro et al. [38] in their work reported that the graphite structure and nodule count can be improved by a well-dosed amount of RE elements in the inoculant composition. This is caused by a large decrement in surface free energy at liquid/graphite interface due to adsorption on graphite. RE elements also favor the graphite shape due to their neutralization effect of excesses of contaminants which cause nodule degeneration. However excess RE metals may cause graphite deterioration, especially in heavy castings where they form microsegregation at grain boundaries, which can be neutralized by proper addition of Bismuth. In another work by Choi et al. [39] it was reported that addition of rare earth elements influences the formation of ferrite at the vicinity of graphite nodules. The addition of RE increases the ferrite thickness with increased casting thickness due to slower cooling rate than that of castings without RE. However, in the case of thicker casting (40 mm), it does not play any significant role on ferrite formation. Also increasing RE up to 0.02% there was a reduction in nodule size and the nodule size increases with increasing RE from 0.02 to 0.03%. However, the graphite nodule size is further decreased when the RE content is again increased to 0.04%. Furthermore, the presence of RE changes nucleation and growth rate by changing the liquid's temperature. The addition of RE appears to increase undercooling, as a result of which its nucleation rate is increased, but the growth rate is decreased. Rare earth elements like lanthanum increases nodule count with increasing content when La to S ratio is as low as 1, hence increases the strength and ductility. However, it does not show any significant development in the nodularity value, which remains almost constant with increasing La amount [21].

3. Case study

3.1 Experimental details

3.1.1 Specimen preparation

In our study to investigate the structure–property relationship, ductile iron test blocks were prepared with different alloying elements and chemical composition by weight percentage is presented in **Table 1**. To investigate the mechanical and tribological properties specimens of 25 × 10 × 5 mm^3 were machined from the test block and then austenitized at 1000°C for 90 min followed by quenched in KNO_3 + $NaNO_3$ (1:1 ratio) at 500°C. The specimens were kept in the salt solution at 500°C for 4 h so as to obtain complete transformation of as-cast matrix. Post quenching the specimens were cooled to room temperature via air cooling, followed by the oxide layer formed on specimen surface was removed by conventional filing and emery paper polishing method.

3.1.2 Vickers hardness and wear testing

The as-cast and austempered specimens are subjected to hardness measurement with the help of a Vickers hardness tester by applying 20 kg load and keeping indenter on the surface for 10 s. Tribological properties were investigated by employing wear test on Ducom TR-208-M1 Ball on plate type wear tester consisting of a 0.4 mm spherical tipped diamond cone indenter of 120° angle. The wear test was conducted at 3 different loads of 10, 20 and 30 N and the travel distance being 7.54 m at a linear velocity of 0.063 m/s. The weights for corresponding specimens before and after test was measured with the help of electronic balance with 0.1 mg accuracy, and prior to the weight measurement specimens were cleaned ultrasonically with acetone before and after the wear took place.

The working principle for ball on plate wear tester is very much similar to that of pin on disc equipment. The only difference between both of the equipment is, the specimen in the pin on disc is in the form of cylindrical pin and the disc on the counterpart rotates against it, where the load is applied on the specimen end. On the

Elements	Wt. %	
	Alloy 1	Alloy 2
C	3.45	3.61
Si	2.07	2.1
Mn	0.15	0.2
S	0.008	0.007
P	0.024	0.022
Cr	0.02	0.03
Ni	0.15	0.47
Cu	—	0.009
Mo	—	0.001
Mg	0.043	0.043
Ce	—	0.004
Fe	Balance	Balance

Table 1.
Chemical composition of specimens in wt.%.

other hand, in case of ball on plate the specimen is fixed on the rotating plate and the indenter being fixed as well as load is applied on this end.

3.1.3 Microstructural and wear morphology characterization

To understand the wear behavior of respective as-cast and austempered specimens' morphological examination was carried out for microstructural aspects. The specimens were initially underwent rough polishing using belt polisher followed by semi finishing polish with 1/0, 2/0, 3/0, 4/0 grades of emery paper and alumina slurry polishing using cloth polisher. The final polishing was carried out using 1 μm diamond paste with the help of cloth polisher. The specimens were observed under microscope in unetched condition to investigate the graphite morphology and distribution in respective a-cast and austempered specimens. And after that each specimen was etched with 2% Nital solution and observed under microscope to obtain the phase characteristics and distribution in the matrix. Additionally the as-cast and austempered specimens subjected to X-ray diffraction investigation for detection of any impurity or precipitation resulted during casting or heat treatment process.

The worn surfaces of respective as-cast and heat treated specimens for both the alloys were observed under optical microscope after the tribology test. The worn surfaces also observed under FESEM and EDAX to understand the wear phenomena and the presence of any foreign element contributing toward the tribological behavior.

3.2 Results and discussion

3.2.1 Morphological characterization

The XRD investigation, **Figure 7**, revealed that there is no trace of any kind of residual phases or precipitation of any impurity present in the final microstructure. It was noticed that only BCC crystallographic planes (110), (200) and (211) are present in the as-cast specimen (**Figure 7(a)**) justifying ferrite and pearlite matrix. On the other hand BCC (110) and (211) and FCC (311) planes were observed for austempered specimens (**Figure 7(b)**) justifying bainitic matrix that is a combination of austenite and ferrite resulted due to the quenching and cooling processes in salt solution and ambient air respectively. The quantitative analysis shows higher carbon content as well as austenite volume fraction for Alloy 2 resulted due to the present of significantly higher amount of Ni as compared to Alloy 1, which act as austenite promoter during the austenitization stage and also helps in retaining the phase during the transformation process to the room temperature microstructure [40].

The microstructures of as-cast and austempered specimens for Alloy 1 and Alloy 2 were illustrated in **Figure 8**. The as-cast matrix of Alloy 1 (**Figure 8(a)**) appeared to have a mixture of ferrite, the white areas around graphite nodules and pearlite, the black area surrounding the white ferrite area in nature. This structure is referred as bull's eye ferritic/pearlitic one with graphite nodules surrounded by ferrite. On the other hand, microstructure of Alloy 2 (**Figure 8(b)**) has fully ferritic in nature surrounding the graphite globules. The difference is attributed to higher amount of Si content which is a strong ferrite promoter [41]. **Figure 8(c)** shows the coarse upper bainitic microstructure resulted by austempered heat treatment process. Quantitative metallographic analysis illustrates that nodularity (roundness of graphite particles) is above 90% that belongs to the Grade I nodularity, for both as-cast and heat treated specimens in both the alloys. The graphite characteristic (nodularity) and distribution (nodule count) for respective alloys in as-cast as well as austempered condition is presented in **Table 2**. It can be observed that Alloy 1 has less no. of nodules than Alloy 2, resulted due to lower Si content and additional Ce

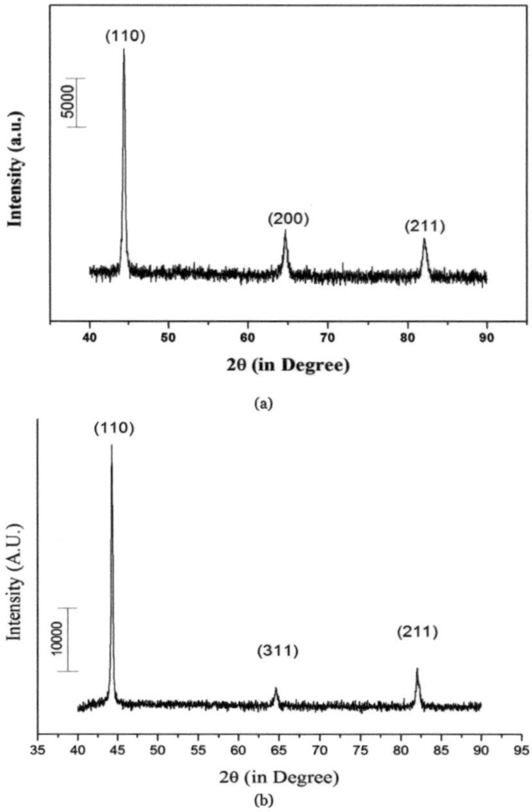

Figure 7.
X-ray diffraction pattern for respective specimens. (a) As-cast and (b) austempered.

as well as higher Ni content in the latter case [42, 43]. The quantitative analysis also reveals the higher bainite volume fraction for Alloy 2 as a consequence of higher Ni content, as was observed from quantitative X-ray diffraction analysis.

3.2.2 Hardness and wear system response

The Vickers hardness value for the as-cast and austempered specimens of both Alloy 1 and Alloy 2 are presented in **Figure 9**. The hardness was found to be higher for austempered specimen with upper bainitic matrix as compared to the as-cast ferritic and ferritic + pearlitic specimens. Alloy 1 has higher hardness in as-cast condition as compared to Alloy 2 which is credited to the amount of pearlite in the matrix, whereas Alloy 2 has only ferritic matrix in nature. On the other hand the case is reverse for austempering treated specimens, i.e., Alloy 2 has higher hardness value than Alloy 1 because of higher amount of Ni in the former resulting combination of higher amount of austenite and eutectoid ferrite in the matrix.

The wear system appears to have resulted in variation of weight loss/gain with variation of applied load for respective as-cast and austempered specimens and decreases continuously for as-cast specimen when load was increased from 10 to 20 N. On the contrary upon increase in load to 30 N a gain was observed which might be due to hardening of surface when operating under repeated no. of cycles. On the other hand, weight gain was observed for austempered specimens when operating under 20 and 30 N. On comparing with hardness values for respective specimens the weight loss at 10 N was observed to be lowest for the specimen with higher hardness value consequence of pearlite in the matrix [44]. The effect of

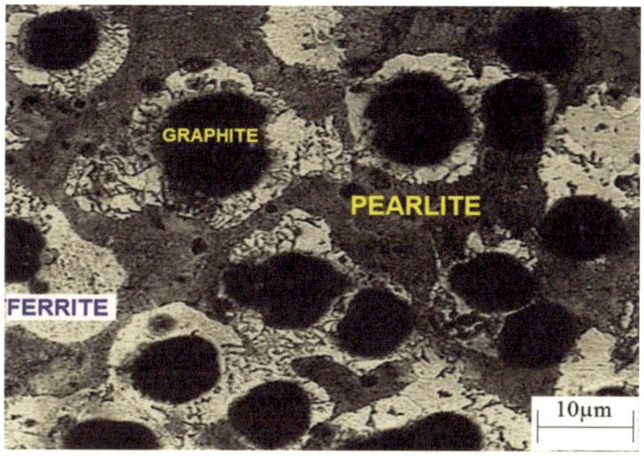

(a)

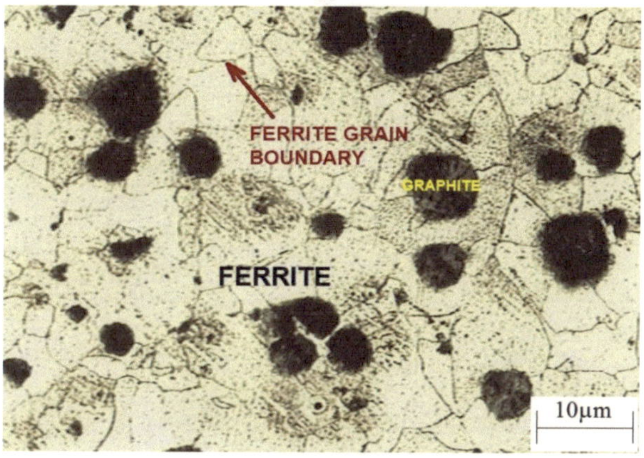

(b)

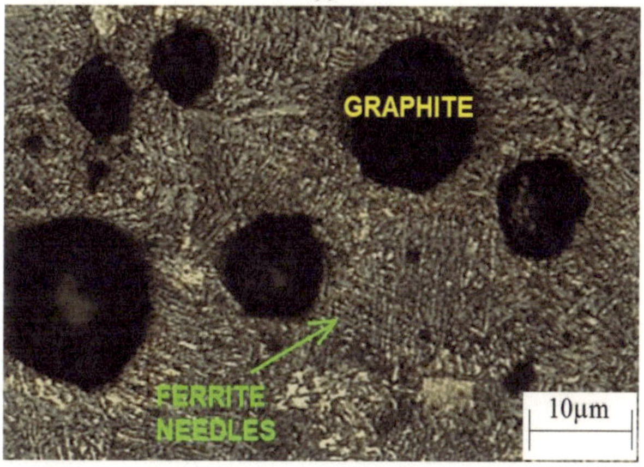

(c)

Figure 8.
Microstructure of as-cast and austempered specimens. (a) As-cast matrix for Alloy 1, (b) as-cast matrix for Alloy 2, and (c) austempered matrix for both Alloy 1 and Alloy 2.

matrix was appeared to have direct effect on the wear rate, where the hardness was higher for bainitic matrix and the wear rate is less as compared to the ferritic and ferritc/pearlitic specimens with lower hardness with higher rate of wear. At 20 N load upper bainitic matrix showed a marginal gain in weight, but as-cast specimen had significantly lose some weight. Further at 30 N again weight gain was observed for both austempered and as-cast specimen.

Alloy	Heat treatment	Nodularity (in %)	Nodule count (mm^{-2})	Phase volume fraction (in %)		Carbon content in austenite
				Austenite	Ferrite	
Alloy 1	As-cast	92	28	—	—	—
Alloy 2		95	33	—	—	—
Alloy 1	Austempered	91	43	70.00%	30.00%	2.5387
Alloy 2		93	50	88.51%	11.49%	2.5941

Table 2.
Microconstituents of as-cast and austempered specimens of respective alloys.

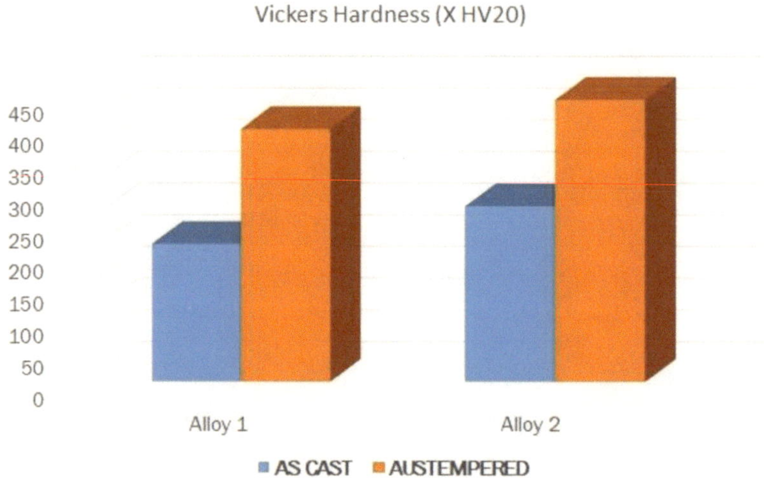

Figure 9.
Vickers hardness for respective as-cast and austempered specimen.

3.2.3 Wear morphology

To understand the wear mechanism involved in due process the worn surfaces for respective as-cast and austempered specimens of both alloys were observed under optical as well as finite element scanning electron microscope and illustrated in **Figures 10** and **11** respectively. The direction of tool travel or wear direction is represented by green arrow for the as-cast and austempered specimens (**Figure 10**). The principal wear mechanism for as-cast and austempered specimens was found to be mostly adhesive wear in nature for all loads. Micro cracks were observed while operating under 10 N for as-cast specimens (**Figure 10(a)**) along with delaminated layer initiated for the soft matrices [45, 46]. Shallow pits and broken wear platelets were also appeared on the surface caused by delamination and work hardening, which was deposited over the graphite nodules. Upon increase in load to 20 N and running for a longer cycles the delaminated layer were observed to be deposited over the graphite nodules (**Figure 10(b)**). The graphite nodules were also observed to expand in transverse direction of wear along with shear deformation [47]. On further increment of load to 30 N (**Figure 10(c)**) the micro cracks were appeared to broaden and surface becomes flat as well as free of wear debris. The wear surface for the austempered specimen under 10 N (**Figure 10(d)**) on the other hand does not have any

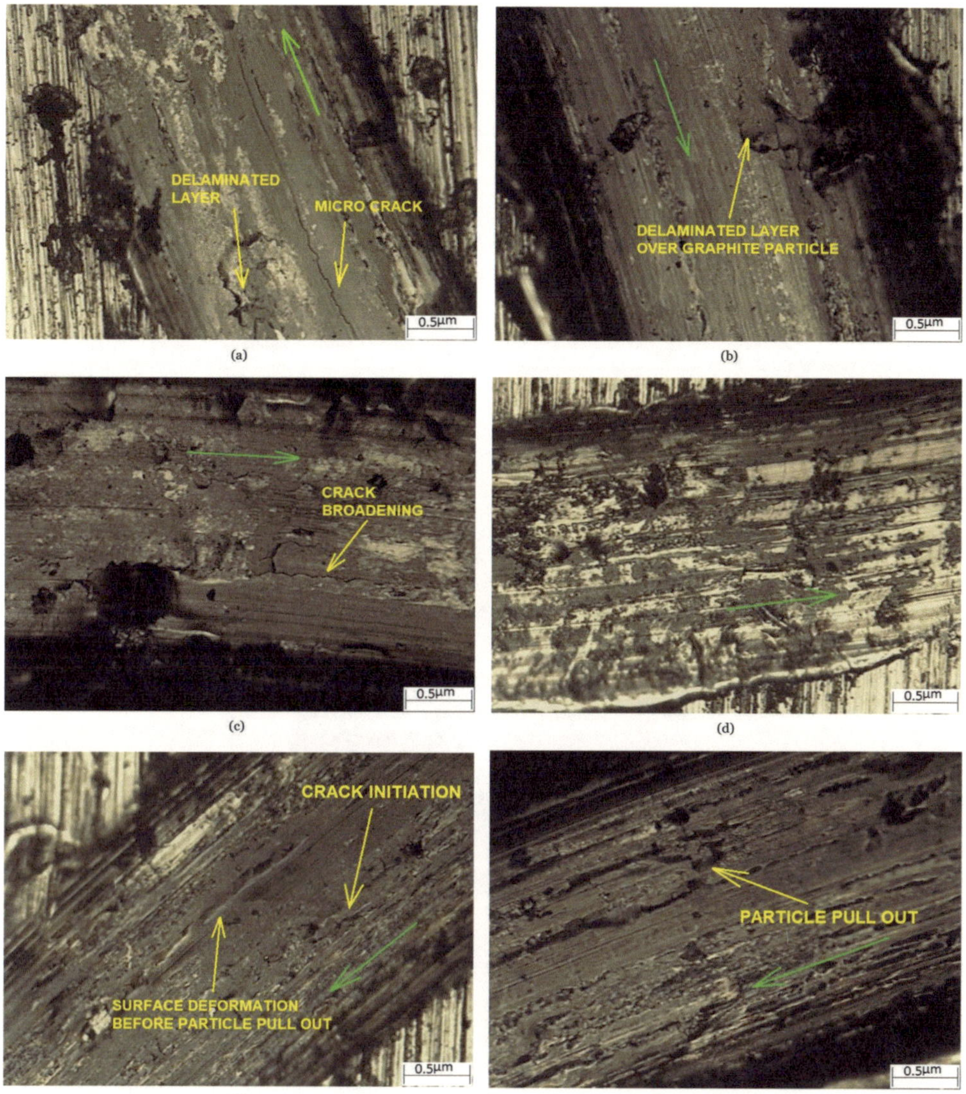

(a) (b) (c) (d) (e) (f)

Figure 10.
Optical microscopy investigation of worn surface of respective specimen at varying load condition. (a) As-cast specimen under 10 N; (b) as-cast specimen under 20 N; (c) as-cast specimen under 30 N; (d) austempered specimen under 10 N; (e) austempered specimen under 20 N and (f) austempered specimen under 30 N.

cracks formed and appeared to be flat in nature with no specific characteristics. The result can be attributed to the hardness of austempered specimen resulting increased wear resistance and work hardening at an early stage. Upon increase in load to 20 N (**Figure 10(e)**) cracks were initiated at the surface as well as small amount of plastic deformation was observed to be involved that might be due to the attention of plastic stage while running under such high load for a large cycle [48, 49]. The plastic deformed feature was observed to be pulled out (**Figure 10(f)**) of the surface upon continuous frictional effect aroused upon increase in load to 30 N. After such a long run the wear surface does not appear to have any crack on it which is probably due to the particle pull out phenomena resulted through plastic deformation, hence the mechanism involved in this case also classified as adhesive wear.

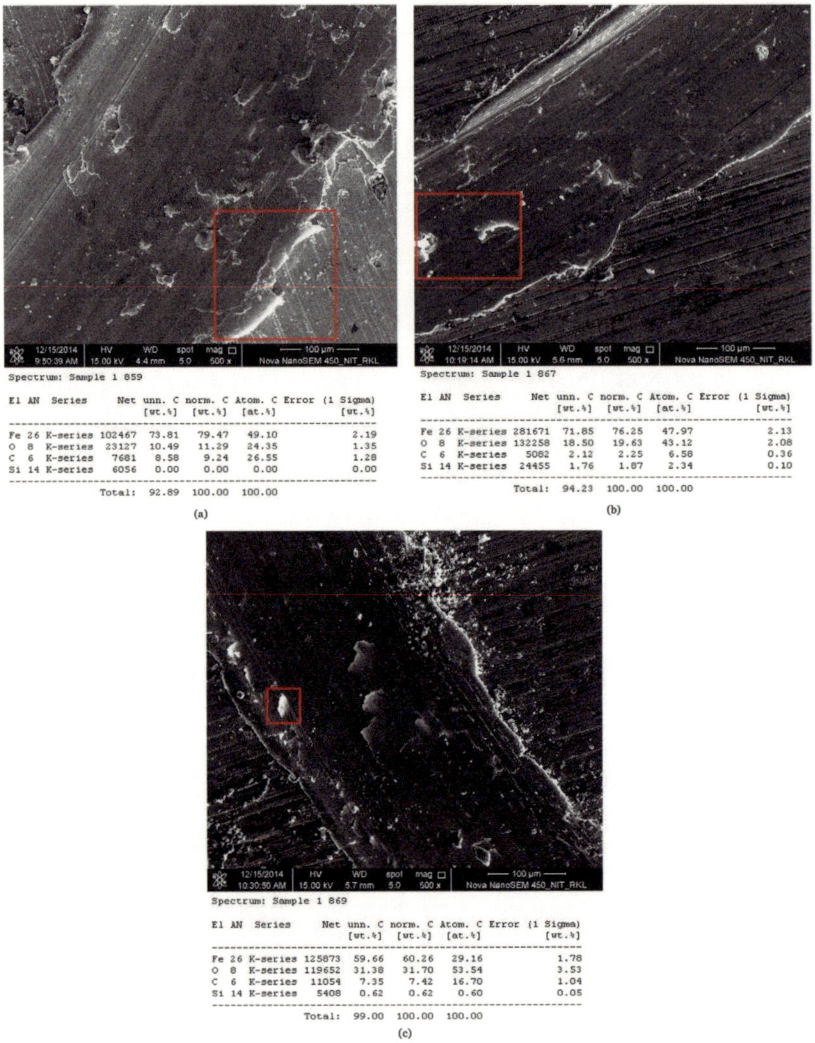

Spectrum: Sample 1 859

El	AN	Series	Net	unn. C [wt.%]	norm. C [wt.%]	Atom. C [at.%]	Error (1 Sigma) [wt.%]
Fe	26	K-series	102467	73.81	79.47	49.10	2.19
O	8	K-series	23127	10.49	11.29	24.35	1.35
C	6	K-series	7681	8.58	9.24	26.55	1.28
Si	14	K-series	6056	0.00	0.00	0.00	0.00
			Total:	92.89	100.00	100.00	

(a)

Spectrum: Sample 1 867

El	AN	Series	Net	unn. C [wt.%]	norm. C [wt.%]	Atom. C [at.%]	Error (1 Sigma) [wt.%]
Fe	26	K-series	281671	71.85	76.25	47.97	2.13
O	8	K-series	132258	18.50	19.63	43.12	2.08
C	6	K-series	5082	2.12	2.25	6.58	0.36
Si	14	K-series	24455	1.76	1.87	2.34	0.10
			Total:	94.23	100.00	100.00	

(b)

Spectrum: Sample 1 869

El	AN	Series	Net	unn. C [wt.%]	norm. C [wt.%]	Atom. C [at.%]	Error (1 Sigma) [wt.%]
Fe	26	K-series	125873	59.66	60.26	29.16	1.78
O	8	K-series	119652	31.38	31.70	53.54	3.53
C	6	K-series	11054	7.35	7.42	16.70	1.04
Si	14	K-series	5408	0.62	0.62	0.60	0.05
			Total:	99.00	100.00	100.00	

(c)

Figure 11.
FESEM and EDAX observation for respective specimens with oxide layers. (a) As-cast specimen under 30N, (b) Austempered at 20N load, (c) Austempered at 30N load.

To investigate the presence of any impurity or foreign contamination the worn surfaces of respective specimens were investigated under NOVA NANOSEM 450, field emission scanning electron microscope, with EDAX analysis feature. The as-cast specimens under 10 N and 20 N did not have any significant characteristics on the surface. However at 30 N load the surface appears to have oxide particles present revealed by the FESEM image (**Figure 11(a)**). On the other hand for austempered specimen the oxide particles were observed for both 20 N (**Figure 11(b)**) and 30 N (**Figure 11(c)**). The oxide particles were observed at different places on the wear ring, and that is responsible for the weight gain as observed in the investigation. The respective areas of oxide presence are shown by red rectangles in respective FESEM images along with the EDAX composition analysis at that particular area. The width and depth of wear track was increased with increase in load as well as responsible for reduction in the crack length for respective specimens. The worn surface features are found to be distinctive for 30 N load as compared to 10 and 20 N. The as-cast specimens appeared to have plowing mark at various places and graphite nodules are also observed to start disrupting from their original shape, confirming adhesive wear mechanism. **Figure 12**

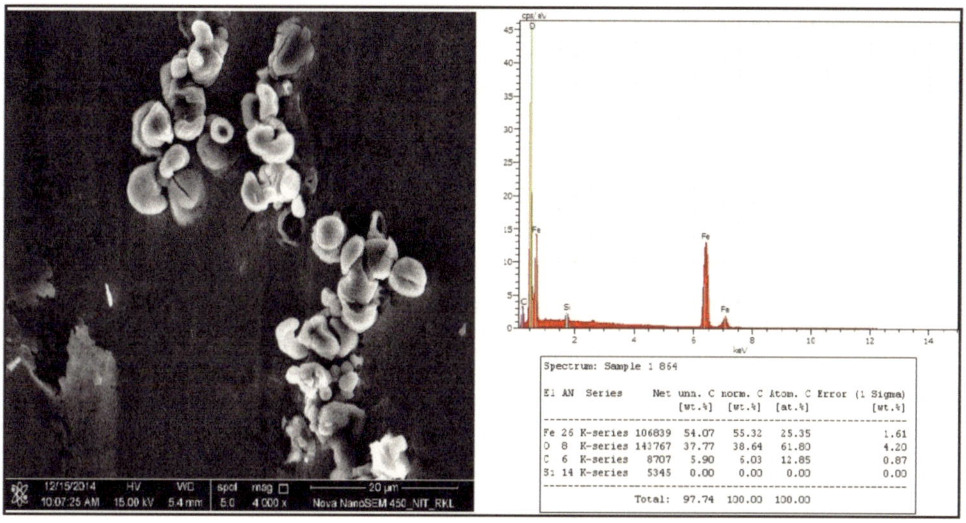

Figure 12.
EDAX of austempered specimen showing oxides when operated under 30 N load.

depicts the EDAX spectrum of austempered specimen under 30 N load showing the highest peak of oxygen along with Fe.

3.3 Conclusion

The wear behavior of as-cast pearlitic/ferritic ductile iron was compared with upper bainitic austempered specimens with different alloy composition. From investigation the following conclusions can be drawn:

1. Austempering at higher temperature leads to transformation of as-cast ferritic and pearlitic/ferritic matrix for both alloys into coarse upper bainitic matrix and increases the hardness.

2. Soft ferritic matrix was less resistant to wear due to lower level of hardness, whereas austempered coarse upper bainitic matrix with higher hardness value was more resistant to wear.

3. A considerable gain in weight was observed for austempered specimen at 20 N load, whereas the normalized specimen does not show any weight loss or weight gain.

4. Adhesive wear phenomenon was observed for specimens with soft ferritic matrix along with micro cracks due to the hardening of surface while operating under high load and kept on increasing with increase in load.

5. The worn surface when subjected to 10 N load did not show any significant features for austempered specimens, but on the contrary fro as-cast soft phase matrices showed plowing micro cracks and delamination features. On the other hand for 30 N loading condition delaminated layer deposition over graphite particles were observed for as-cast specimen and austempered specimen showed particle pull phenomena along with breaking of wear continuity. Under all the loads the wear phenomenon was observed to be adhesive type wear.

Acronyms and abbreviations

SGCI	spheroidal graphite cast iron
DCI	ductile cast iron
ADI	austempered ductile iron

Author details

Ranjan Kumar Behera[1*], Sudipta Sen[2] and Subash Chandra Mishra[2]

1 Accendere Knowledge Management Services Pvt Ltd, New Delhi, India

2 Department of Metallurgical and Materials Engineering, National Institute of Technology Rourkela, Rourkela, India

*Address all correspondence to: drranjannitr@gmail.com

References

[1] Putatunda SK. Influence of austempering temperature on fracture toughness of a low manganese austempered ductile iron (ADI). Materials and Manufacturing Processes. 2001;**16**(2):245-263

[2] Rebasa N, Dommarco R, Sikora J. Wear resistance of high nodule count ductile iron. Wear. 2002;**253**(7-8):855-861

[3] Motu1. Effect of Heat Treatment Prodedures on Microstructureand Mechanical Properties of Nodular Iron. 2009. pp. 1-97. Available from: http://ethesis.nitrkl.ac.in/1501/1/motu-thesis.pdf

[4] Abdullah B, Alias SK, Jaffar A, Rashid AA, Ramli A. Mechanical properties and microstructure analysis of 0.5% niobium alloyed ductile iron under austempered process in salt bath treatment. 2010 International Conference on Mechanical and Electrical Technology (ICMET). 2010. pp. 610-614

[5] Shamanth DS, Murthy KN. Tensile strength and mechancal properties evaluation of permanent mould ductle iron subjected to novel two step austempering process. International Journal of Engineering Research and Applications (IJERA). 2012;**2**(3):2621-2627

[6] Zammit a, Abela S, Wagner L, Mhaede M, Grech M. Tribological behaviour of shot peened Cu-Ni austempered ductile iron. Wear. 2013;**302**(1-2):829-836. Available from:. DOI: 10.1016/j.wear.2012.12.027

[7] Masud L, Martínez R, Simison S, Boeri R. Embrittlement of austempered ductile iron on contact with water-testing under applied potential. Journal of Materials Science. 2003;**38**(13):2971-2977

[8] Elsayed AH, Megahed MM, Sadek AA, Abouelela KM. Fracture toughness characterization of austempered ductile iron produced using both conventional and two-step austempering processes. Materials and Design. 2009;**30**(6):1866-1877. Available from:. DOI: 10.1016/j.matdes.2008.09.013

[9] Hsu CH, Chen ML. Corrosion behavior of nickel alloyed and austempered ductile irons in 3.5% sodium chloride. Corrosion Science. 2010;**52**(9):2945-2949. DOI: 10.1016/j.corsci.2010.05.006

[10] Sohi MH, Ahmadabadi MN, Vahdat AB. The role of austempering parameters on the structure and mechanical properties of heavy section ADI. Journal of Materials Processing Technology. 2004;**153-154**(1-3):203-208

[11] Balachandran G, Vadiraj A, Kamaraj M, Kazuya E. Mechanical and wear behavior of alloyed gray cast iron in the quenched and tempered and austempered conditions. Materials and Design. 2011;**32**(7):4042-4049. DOI: 10.1016/j.matdes.2011.03.054

[12] Putatunda SK. Comparison of the mechanical properties of austempered ductile cast iron (ADI) processed by conventional and step-down austempering process. Materials and Manufacturing Processes. 2010;**25**(8):749-757. Available from: http://www.scopus.com/inward/record.url?eid=2-s2.0-77956504093&partnerID=40&md5=c72bebbdefb64febdae0424053023059

[13] Krawiec H, Stypuła B, Stoch J, Mikołajczyk M. Corrosion behaviour and structure of the surface layer formed on austempered ductile iron in concentrated sulphuric acid. Corrosion Science. 2006;**48**(3):595-607

[14] Zandira M, Boutorabi SMA. Fracture characteristics of austempered spheroidal graphite aluminum cast irons. Journal of Iron and Steel Research, International. 2010;**17**(2):31-35. DOI: 10.1016/S1006-706X(10)60055-6

[15] Kayali Y, Büyüksağiş a, Yalçin Y. Corrosion behavior of boro-tempered ductile iron. Protection of Metals and Physical Chemistry of Surfaces. 2010;**46**(3):345-349

[16] Sugishita J, Fujiyoshi S. The effect of cast iron graphite on friction and wear performance III: The lubricating effect of graphite under rolling-sliding contacts. Wear. 1982;**77**(2):181-193. DOI: 10.1016/0043-1648(82)90102-8

[17] Zimba J, Samandi M, Yu D, Chandra T, Navara E, Simbi DJ. Un-lubricated sliding wear performance of unalloyed austempered ductile iron under high contact stresses. Materials and Design. 2004;**25**(5):431-438

[18] Bai Y, Luan Y, Song N, Kang X, Li D, Li Y. Chemical compositions, microstructure and mechanical properties of roll core used ductile iron in centrifugal casting composite rolls. Journal of Materials Science and Technology. 2012;**28**(9):853-858. DOI: 10.1016/S1005-0302(12)60142-X

[19] Ingole PPM, Awate AU, PSV S. Effect of basic chemical element in sgi (ductile iron). International Journal of Engineering Research & Technology (IJERT). 2012;**1**(7):1-7

[20] Neri MA, Carreño C. Effect of copper content on the microstructure and mechanical properties of a modified nodular iron. Materials Characterization. 2003;**51**(4):219-224. Available from: http://linkinghub.elsevier.com/retrieve/pii/S1044580303001487

[21] Available from: http://www.ductile.org/didata/Section3/3part2.htm#Manganese Copper Nickel Phosphorus

[22] Available from: http://www.nrc.gov/waste/spent-fuel-storage/diagram-typical-trans-cask-system.doc

[23] Available from: http://www.ductile.org/didata/Section2/figures/pfig2_8.htm

[24] Mochizuki S, Matsushita H. Physical properties of nodular cast iron for shipping containers and safety analysis by fracture mechanics. Nuclear Engineering and Design. 1986;**94**(3):309-316

[25] Droste B. A fracture mechanics safety concept to assess the impact behavior of ductile cast iron containers for shipping and storage of radioactive materials, Nuclear Engineering and Design. 1994;**150**:357-366

[26] Sorenson KB, Salzbrenner RJ. Quality assurance aspects in using ductile cast iron for transportation casks. United States: University of Arizona Nuclear Engineering Dept;1988

[27] Teng TL, Chu YA, Chang FA, Chin HS, Lee MC. The dynamic analysis of nuclear waste cask under impact loading. Annals of Nuclear Energy. 2003;**30**(14):1473-1485

[28] Jakšić N, Nilsson KF. Finite element modelling of the one meter drop test on a steel bar for the CASTOR cask. Nuclear Engineering and Design. 2009;**239**(2):201-213

[29] Baer W, Bösel D, Eberle A, Klingbeil D. Determination of dynamic crack resistance of ductile cast iron using the compliance ratio key curve method. Engineering Fracture Mechanics. 2010;**77**(2):374-384. DOI: 10.1016/j.engfracmech.2009.06.015

[30] Baer W, Wossidlo P, Abbasi B, Cassau M, Häcker R, Kossert R. Large

scale testing and statistical analysis of dynamic fracture toughness of ductile cast iron. Engineering Fracture Mechanics. 2009;**76**(8):1024-1036. DOI: 10.1016/j.engfracmech.2009.01.005

[31] Nilsson KF, Vokál V. Analysis of ductile cast iron tensile tests to relate ductility variation to casting defects and material microstructure. Materials Science and Engineering A. 2009;**502**(1-2):54-63

[32] Zencker U, Zeisler P, Droste B. Dynamic fracture mechanics assessments for cubic ductile cast iron containers. International Journal of Radioactive Materials Transport. 2000;**11**:113-118

[33] Available from: http://www.ductile.org/didata/Section2/figures/pfig2_5.htm

[34] Uma TR, Simha JB, Murthy KN. Influence of nickel on mechanical and slurry erosive wear behaviour of permanent moulded toughened austempered ductile iron. Wear. 2011;**271**(9-10):1378-1384. DOI: 10.1016/j.wear.2010.12.050

[35] Fatahalla N, AbuElEzz A, Semeida M. C, Si and Ni as alloying elements to vary carbon equivalent of austenitic ductile cast iron: Microstructure and mechanical properties. Materials Science and Engineering A. 2009;**504**(1-2):81-89. Available from: http://linkinghub.elsevier.com/retrieve/pii/S0921509308012161

[36] Zencker U, Weber M, Kovacs O, Qiao L, Droste B. Application limits of low-ductile cast iron. Proceedings of the 15th International Symposium on the Packaging and Transportation of Radioactive Materials PATRAM 2007 October 21-26. Miami, Florida, USA; 2007. pp. 1-8

[37] Jafar KA, Behnam AA. Influence of mold preheating and silicon content on microstructure and casting properties of ductile iron in permanent mold. Journal of Iron and Steel Research, International. 2011;**18**(3):34-39. DOI: 10.1016/S1006-706X(11)60034-4

[38] Ferro P, Fabrizi a, Cervo R, Carollo C. Effect of inoculant containing rare earth metals and bismuth on microstructure and mechanical properties of heavy-section near-eutectic ductile iron castings. Journal of Materials Processing Technology. 2013;**213**(9):1601-1608. DOI: 10.1016/j.jmatprotec.2013.03.012

[39] Choi JO, Kim JY, Choi CO, Kim JK, Rohatgi PK. Effect of rare earth element on microstructure formation and mechanical properties of thin wall ductile iron castings. Materials Science and Engineering A. 2004;**383**(2):323-333

[40] Kilicli V, Erdogan M. The nature of the tensile fracture in austempered ductile iron with dual matrix microstructure. Journal of Materials Engineering and Performance. 2010;**19**(1):142-149

[41] Cui J, Chen L. Microstructure and abrasive wear resistance of an alloyed ductile iron subjected to deep cryogenic and austempering treatments. Journal of Materials Science and Technology. 2017;**33**(12):1549-1554. Available from: https://linkinghub.elsevier.com/retrieve/pii/S1005030217302086

[42] Bedolla-Jacuinde A, Guerra FV, Rainforth M, Mejía I, Maldonado C. Sliding wear behavior of austempered ductile iron microalloyed with boron. Wear. 2015;**330-331**:23-31. Available from: https://linkinghub.elsevier.com/retrieve/pii/S0043164815000113

[43] Zhang H, Wu Y, Li Q, Hong X. Mechanical properties and rolling-sliding wear performance of dual phase austempered ductile iron as potential metro wheel material. Wear. 2018; **406-407**:156-165. Available from:

https://linkinghub.elsevier.com/retrieve/pii/S0043164818300759

[44] Polajnar M, Kalin M, Thorbjornsson I, Thorgrimsson JT, Valle N, Botor-Probierz A. Friction and wear performance of functionally graded ductile iron for brake pads. Wear. 2017;**382-383**:85-94. Available from: https://linkinghub.elsevier.com/retrieve/pii/S0043164816308201

[45] Sun X, Wang Y, Li DY, Wang G. Modification of carbidic austempered ductile iron with nano ceria for improved mechanical properties and abrasive wear resistance. Wear. 2013;**301**(1-2):116-121. Available from: https://linkinghub.elsevier.com/retrieve/pii/S0043164812004462

[46] Cao HT, Dong XP, Pan Z, Wu XW, Huang QW, Pei YT. Surface alloying of high-vanadium high-speed steel on ductile iron using plasma transferred arc technique: Microstructure and wear properties. Materials and Design. 2016;**100**:223-234. Available from: https://linkinghub.elsevier.com/retrieve/pii/S0264127516303987

[47] Arabi Jeshvaghani R, Jaberzadeh M, Zohdi H, Shamanian M. Microstructural study and wear behavior of ductile iron surface alloyed by Inconel 617. Materials and Design. 2014 Feb;**54**:491-497. Available from: https://linkinghub.elsevier.com/retrieve/pii/S0261306913008030

[48] Zhang N, Zhang J, Lu L, Zhang M, Zeng D, Song Q. Wear and friction behavior of austempered ductile iron as railway wheel material. Materials and Design. 2016;**89**:815-822. Available from: https://linkinghub.elsevier.com/retrieve/pii/S0264127515306158

[49] Samaddar S, Das T, Chowdhury AK, Singh M. Manufacturing of engineering components with austempered ductile iron—A review. Materials today: Proceedings. 2018;**5**(11):25615-25624. Available from: https://linkinghub.elsevier.com/retrieve/pii/S2214785318327366

Chapter 7

Field Ion Microscopy of Radiation Effects in Metallic Materials

Vladimir Alexandrovich Ivchenko

Abstract

Radiation effects in metallic materials were investigated using the field ion microscopy technique. Modes of radiation exposure for development of amorphized states in subsurface regions of platinum are determined. It is found that the phenomenon of metal amorphization in the subsurface regions occurs up to a sample depth of 12 nm under an increase in the fluence to 10^{18} ions/cm^2 and the above irradiation energies. The method of field ion microscopy was used to determine the threshold of nanopore formation in ion-implanted platinum. Experimental results on atomic-spatial investigation of radiative defect formation in surface layers of materials, initiated by neutron bombardment (of Pt, $E > 0.1$ MeV) and ion implantation (in Cu_3Au: $E = 40$ keV, $F = 10^{16}$ ion/m^2, $j = 10^{-3}$ A/cm^2), are considered. Mechanisms of the radiation-induced development of nanostructures in subsurface metal regions have been analyzed based on field ion microscopy data. It is concluded that the modification of near-surface metal regions on a nanometer scale as a result of the interaction with Ar^+ ion beams proceeds by several mechanisms.

Keywords: field ion microscopy, nanostructure states, metals, alloys, a radiating irradiation, interaction of ions with matter, structure modification

1. Introduction

One high-priority area in condensed matter physics is related to work in the field of production of new materials via severe exposure. Such types of exposure include irradiation techniques.

Experimental studies of changes in the metal and alloy structure after radiation exposure [1–6] showed that interaction between accelerated charged ions and the substance initiates formation of amorphous, nanomicrocrystalline, and submicrocrystalline states in the subsurface regions.

One important problem in radiation material science is the determination of the structural state and phase composition of subsurface layers of materials subjected to irradiation. One topical problem consists in studies of interaction between charged particle beams and the material surface in the subsurface regions (at a distance of ~1–100 nm from the irradiated surface). Therefore, in this work, precision studies of changes in the actual structure surface of atomic layers of metals were carried out using the field ion microscopy (FIM) technique. The options of FIM allow studying the subsurface regions of irradiated materials using controlled removal of atoms from the surface and thus analyzing the sample structure in the course of layer-by-layer field evaporation of atoms. Visualization of an atomically pure and atomically smooth surface of the research object at cryogenic temperatures allows obtaining quantitative results of changes in atom positions in the crystal lattice due to ion implantation doping of positive argon atoms.

Intensely developing studies of changes in the structure of metals and alloys after interaction between particle beams and the surface show that ionic treatment results in the formation of specific condensed states as well as unique strength and physical properties in materials that cannot be obtained using conventional methods [1, 3–5, 7, 8]. It is known that pore swelling in materials under irradiation causes shortening of the operational life of products and drastic deterioration of physical and mechanic properties of materials. Therefore, studies of the appearance of nanopores on the surface and in the subsurface layer of the studied objects under exposure to radiation on an atomic scale appear to be of current interest.

In this work, the results of modification of the surface and subsurface region of pure metals (Pt) after irradiation by medium energy-charged particles (up to 30 keV) within a nanometer deep surface region are shown. The method of field ion microscopy (FIM) allows studying experimentally structural states appearing under such exposure.

Investigating interaction mechanisms of accelerated particles with matter and studying the atomic rearrangement and, as a consequence, formation of crystal lattice defects and changing the phase state of the material are important tasks in radiation physics of solids.

This work is devoted to studying the spatial distribution of radiation damages, particularly vacancies and their complexes in the bulk fcc metals exposed to different irradiation (by neutron and ion beams).

Imitation of the neutron irradiation of high fluence with the help of positively charged ion beams allows one to solve the problem of analog simulation of radiation generated on one setup when replacing it by the radiation generated on another setup.

The results of our previous investigations [9, 10] of the surface structure of metals, carried out using the field ion microscopy (FIM) method, showed that the irradiation of metal targets by charged particles (argon ions) with moderate energies (up to 40 keV) led to nanostructurization of a near-surface region of pure metals. Using the FIM technique, it is possible to study the subsurface layers of irradiated materials by controlled removal of atoms from the sample surface and analyze its structure in the course of layer-by-layer field evaporation of atoms. Sequential imaging of the atomically clean surface of a sample at cryogenic temperatures provides quantitative information on the positions of individual atoms and atomic groups and their changes as a result of the implantation of positive argon ions. Thus, it is possible to study the atomic structure of radiation-induced defects in the crystalline lattice, determine their distribution in the volume, estimate the thickness of a modified subsurface layer, etc.

This work presents an attempt to analyze the effect of ion implantation on the atomic structure of platinum depending on the variable parameters (including the ion energy, ion current density, and dose) of irradiation with beams of accelerated argon ions. The aim was to elucidate the mechanisms of formation of nanostructural states in the ion-modified subsurface region of a target. Investigation of the laws of these phenomena is necessary for developing methods of controlled modification of the physical properties of materials, which is an important area in advanced nanotechnologies. In addition, the interaction of ions with substances provides a basis for methods of diagnostics of the initial materials and changes to them caused by ion irradiation.

2. Experimental part

2.1 FIM technique

The most powerful and modern methods of investigation of conducting and semiconductor materials, which can be used to directly study the crystal lattice of solids with atomic-spatial resolution, include field ion microscopy (FIM) and various modifications of the atomic probes of the field ion microscope.

Despite the fact that field ion microscopy (FIM) has entered its sixth decade, it is still the only technique in microscopy capable of providing direct observation of individual atoms as elements that make up the structure of the sample during a conventional experiment. Using FIM it is possible to select a single atom in the image for mass spectrometric identification (atomic probe FIM methods) to carry out in situ experiments with individual atoms deposited on the surface and to reconstruct in volume the structural and chemical composition of the sample by means of controlled layer-by-layer removal of surface atoms by an electric field at cryogenic temperatures. It is the latter circumstance that distinguishes direct methods of PIM from other structurally sensitive, atomic-resolution, but indirect, methods of microscopic study of materials.

The operating principle of a field ion microscope is based on projective ion imaging of the sample surface and is controlled by the device design [11]. The ion image of the sample vertex surface [$r \sim (3–5) \times 10^6$ cm] (**Figure 1**) is formed on the microscope screen by positive ions of an imaging gas (usually He, Ne) by inducing a strong electric field over the emitter surface ($\sim 5 \times 10^8$ V/cm) at a corresponding

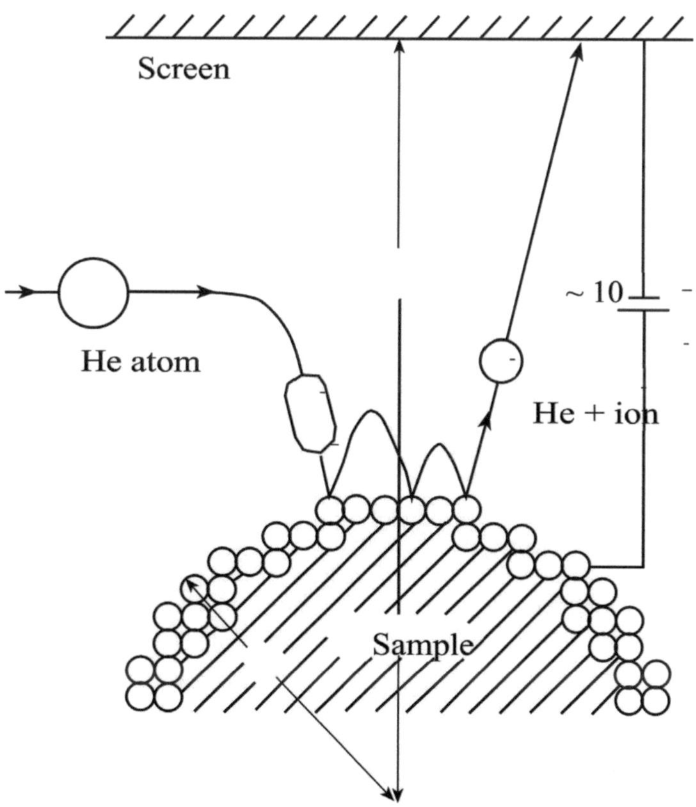

Figure 1.
Schematic representation of ion imaging of the metal surface ($r = 3 \times 10^{-6}$ cm, $R = 6$ cm, sample $T = 78$ K).

potential difference between the sample and shield. The magnification reaches several million diameters, and the resolution is 0.2–0.3 nm. The residual gas pressure in the microscope chamber can be in the range of 10^{-4} to 10^{-8} Pa, depending on the problems of study.

As an imaging gas, an inert gas is used, whose operating pressure is usually 10^{-3}–10^{-4} Pa and is determined by the distance between the tip and fluorescent screen, which is comparable to the mean range of imaging gas ions. Ionization of imaging gas atoms occurs at a certain electric field strength and is observed at a distance no closer than the critical one from the tip end surface, ~0.5 nm. The ionization probability of atoms of the imaging gas is much higher above protruding surface atoms which are arranged at step knee points and are approximately estimated by the Wentzel–Kramers–Brillouin (WKB) relation [12]. Such an event becomes probable owing to electron tunneling through the potential barrier due to the tunneling effect.

Therefore, on the screen of the microscope, an image is observed (**Figure 2**), on the one hand, representing the contrast of the surface of the tip of the sample from the atoms in the fracture steps, and on the other—the stereographic projection of the crystal under study. The circular contour lines on the ion image are the edges of the corresponding families of crystallographic planes of certain directions. Neighboring rings (from any family of concentric rings) are images of parallel atomic layers. The distance between the rings corresponds to the interplanar for a given crystallographic direction. In the terminology of ion micropattern, this distance is called the height of a step of the crystal lattice. The rings themselves on the ion micropattern, as a rule, consist of separate bright points, which represent images of surface atoms located in the positions of atoms in the fractures of the steps.

Imaging of the ion microscopic picture of the sample surface by radial projection is directly associated with the possibility of preparing the emitter tip shaped as an almost hemispherical atomically smooth surface. This is achieved using field evaporation, which occurs only at a sufficiently high electric field strength over the emitter surface. In such a field, atoms protruding from the surface will "evaporate" in the form of positive ions. Field evaporation is a self-controlled process of tip apex "polishing," since local electric field enhancement near sharp edges and over bumps results in their preferential "evaporation" (removal) by the field. As a result, a tip apex surface ideally smooth on the atomic scale is formed. Furthermore, this surface

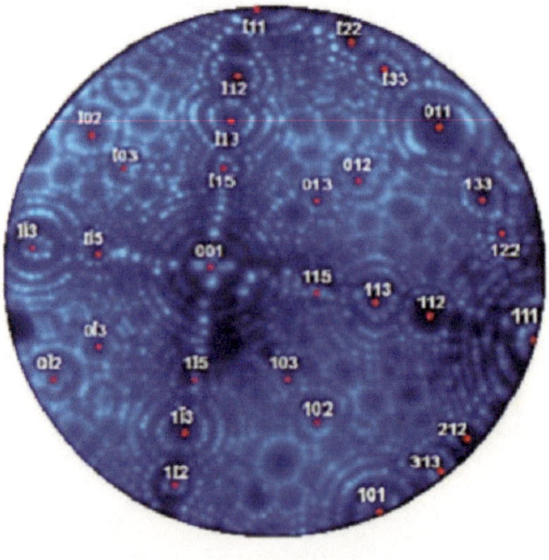

Figure 2.
Neon image of Pt single crystal, U = 10 kV.

permanently visualized during field evaporation is atomically clean. Continuous recording of such a surface using photo, video, or movie cameras during controlled layer-by-layer removal of surface atoms makes it possible to analyze the crystal structure of the object of study in the material volume.

2.2 Production of sharp samples

We used platinum with a purity of 99.99% as an object of irradiation. Using electrochemical polishing of wire fragments, needle-like emitters with a curvature radius of 30–50 nm near the apex point were manufactured. The field emitters certified for ion implantation have an atomically smooth surface of the vertex prepared by field evaporating surface atoms in situ. Irradiation of the needle-like samples certified by FIM was performed by Ar^+ beams with an energy of 30 keV, fluence $F = 10^{16}$ ion/cm^2, and ion density current $j = 150$ μA/cm^2 ($T = 70°$) or $j = 200$ μA/cm^2 ($T = 200°C$). The bombardment was carried out in the direction parallel to the axis of the emitter. Being certified in advance, the implanted samples were placed in FIM once again for studying the subsurface volume of the alloy.

In the study of fast neutron irradiated samples, the tip was made from irradiated billets. Billets of Pt consisted of wire segments with a diameter of 0.2 mm and a length of 20 mm. Billets were irradiated in the IBB-2 M reactor at T = 310 K for 1 h to fast neutron fluences (with E > 0.1 MeV) $F = 6.7 \times 10^{21}$ m^{-2} and $F = 3.5 \times 10^{22}$ m^{-2}. In the field ion microscope, we studied tip samples with a radius of curvature at the top of 10–30 nm. Ion images of the surface were recorded by a photo or video camera with controlled removal of atomic layers. Then the structural state of the alloy in the volume was analyzed.

A field ion microscope was equipped with a micro channel ion–electron converter enhancing the brightness of surface microimages by 10^4 times. The refrigerant agent was generally liquid nitrogen ($T = 78$ K); the imaging gas was spectrally pure neon.

3. Results

3.1 Creation of an amorphized state on the surface of metal materials

The aim of this work is to determine the modes of radiation exposure of Ar^+ beams accelerated to 30 keV for development of amorphized states in subsurface metal regions (Pt). Earlier, the FIM technique was used in [5] to establish the effect of partial amorphization in subsurface regions of the Cu_3Au alloy.

An atomically smooth surface of the emitter tip for the further irradiation was obtained in situ in the course of field evaporation of surface atoms. Ion images of evaluated field emitters registered a practically perfect ring pattern of the pure metal single crystals pointing to the virtual absence of structural defects (**Figure 3a**). Which was shown by analysis of the ion pattern of terns manifested irregularities in the ring patterns of the crystal face images. It is such irregularities in the ring ion pattern that allow registering defects in the perfect crystal structure and determining the patterns of any defects appearing in the material as a result of exposure. In this case, changes in the ion pattern of irradiated platinum observed in the layer at a depth of 1.5 nm from the irradiated surface as compared to the pattern of the initial evaluated Pt indicate the presence of a block nano-size structure in the subsurface regions of the material [9] (**Figure 3b**).

Analysis of the ion pattern of atom positions in nanoblocks (**Figure 3b**) clearly indicates that atoms actually occupy their sites in the crystal lattice of the material, although the blocks themselves are nonoriented.

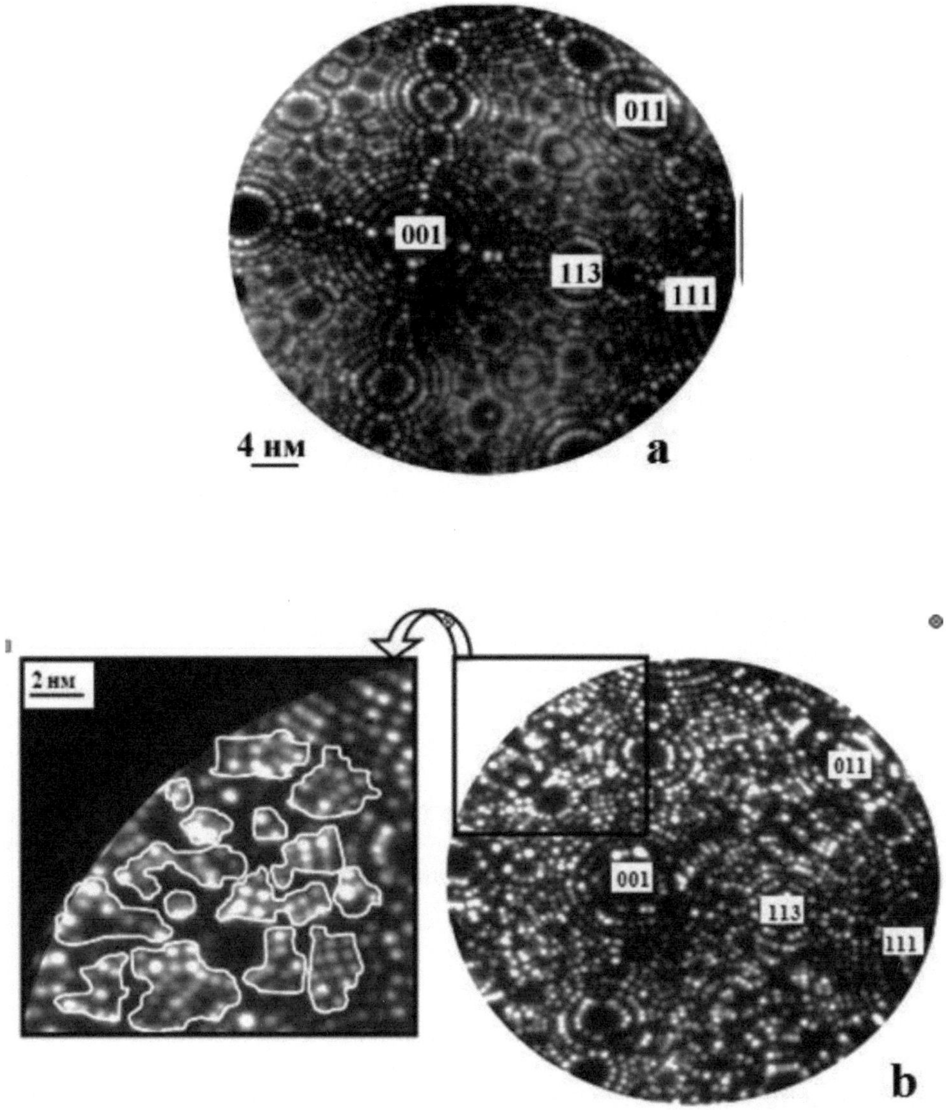

Figure 3.
Neon images of Pt: (a) ion pattern of the evaluated crystal; (b) ion pattern of the surface after irradiation by Ar^+ *with* $F = 10^{16}$ *ions/cm*^{2} *(T = 70°C).*

The effect of the formation of a block nanocrystalline structure (with block size 1–5 nm) is observed in the near-surface volume with a depth of at least 20 nm from the irradiated surface as a result of irradiation to a higher fluence ($F = 10^{17}$ ions/cm^2, **Figure 4**).

Studies of the corresponding experimental data allowed determining the lateral and longitudinal dimensions of nanocrystalline blocks and the width of the boundary region between nanoblocks. The estimated width of the boundary region varied from 0.4 to 0.8 nm at different parts of the nanoblock boundaries in ion irradiated platinum [13].

The ion pattern of the irradiated platinum surface manifests an image typical for grain boundaries and packing defects [14] for practically all micropattern faces (**Figure 4**). This means that at $F = 10^{17}$ ions/cm^2, the mechanism of formation of the nanoblock structure in the body of the material changes.

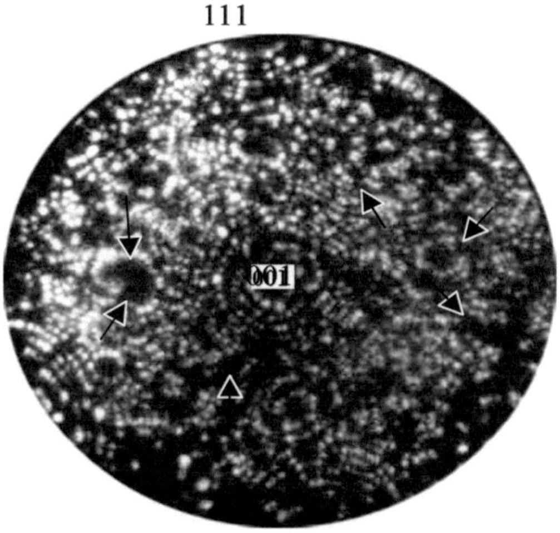

Figure 4.
Neon image of Pt after irradiation by Ar^+ with F = 10^{17} ions/cm^2 (T = 200°C). Typical ion pattern of nanoblock boundaries and defects is shown by arrows.

Radiation exposure of pure metals with E = 30 keV under variation of the fluence of charged argon ion beams by two orders of magnitude (10^{16}–10^{18} ions/cm^2) produces a significant effect on the kinetics of defect formation in the subsurface regions of irradiated materials. The ion image of the irradiated surface of platinum with a fluence of 10^{18} ion/cm^2 is given (**Figure 5**).

Based on the contrast of the micro images of the atomic-pure surface of platinum in the analysis of the near-surface volume of the material in the process of controlled removal of atomic layers, it is obvious that with an increase in the fluence to 10^{18} ion/cm^2, the phase state of the metal practically becomes amorphous. The proof is the structureless arrangement of atoms in the near-surface layers. The analog of the observed ionic contrast corresponds to the ionic contrast of amorphous materials obtained by ultrafast cooling. According to our estimates amorphization of pure metal (Pt) occurs in the subsurface volume with a depth of 12 nm from the irradiated surface.

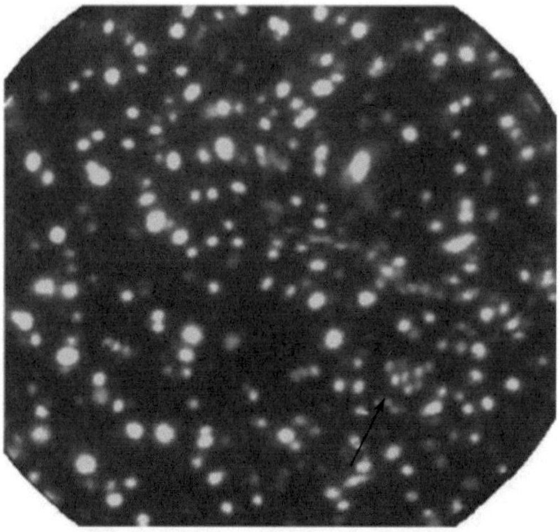

Figure 5.
Ion pattern of Pt surface after irradiation by Ar^+ with F = 10^{18} ions/cm^2 (T = 300°C) (the arrow denotes the region of the crystalline state of the metal).

Thus, the modes of radiation exposure in the creation of amorphized states in the near-surface volume of platinum are determined.

It is shown that the radiation effect on pure metals with E = 30 keV when the fluence of charged argon ion beams changes by two orders of magnitude (from 10^{16} to 10^{18} ion/cm^2) significantly affects the kinetics of defect formation in the near-surface volume of irradiated materials.

It is found that the phenomenon of metal amorphization in the subsurface regions of Pt occurs up to the sample depth of 12 nm under an increase in the fluence to 10^{18} ions/cm^2 and the above irradiation energies. The amorphized phase regions are retained in the metal at a depth of at least 60 nm.

3.2 The threshold of nanopore formation in ion-implanted platinum

The formation of nanopores in metals after their interaction with charged beams of Ar^+ ions was experimentally studied. The task was to analyze the distribution of nanopores in the near-surface volume by size depending on the distance from the irradiated surface and the irradiation fluence. Set the threshold for the formation of nanopores; determine the optimal modes of radiation exposure to obtain nano-structured near-surface volumes in irradiated metals and alloys.

Ionic images of evaluated field emitters registered a regular ring pattern of the pure metal single crystal surface pointing to the virtual absence of structural defects (**Figure 6**). In the microphotograph, faces {111}, {110}, and {113} are also registered, which are characteristic for crystals with an fcc lattice, in addition to the {001} face.

The ring contour lines in ionic images are edges of the corresponding families of crystallographic planes of certain directions. Neighboring rings (from any family of concentric rings) are images of parallel atomic layers. The rings themselves in an ionic microimage generally consist of separate bright points corresponding to images of surface atoms located at the sites of atoms in the step kinks.

During the study of the crystal structure of platinum irradiated to F = 10^{16}–10^{18} ion/cm^2 with an energy of 30 keV and j = 200 μA/cm^2, ion contrast of nanopores was found in the near-surface volume. Such contrast was registered at fluence 10^{17} ion/cm^2.

The contrast of nanopores in Pt was recorded after removal of several atomic layers from the surface of the irradiated metal (**Figure 7a**). This ionic contrast of nanopores was recorded as the contrast of "craters" at the time of removal of the last atomic layer before the appearance of nanopores. During the field evaporation

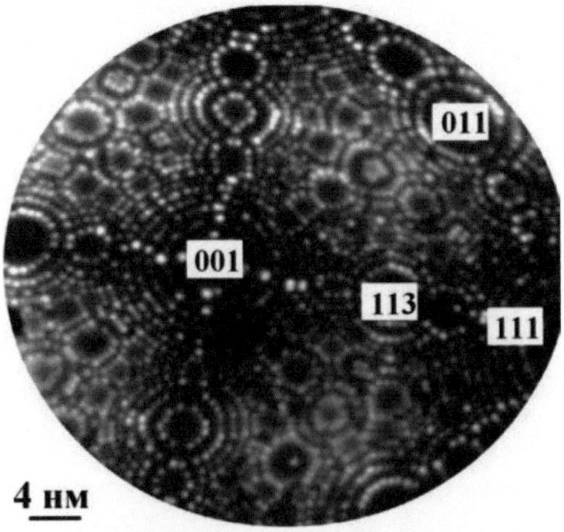

Figure 6.
Neon image of validated Pt single crystal prepared for ionic irradiation.

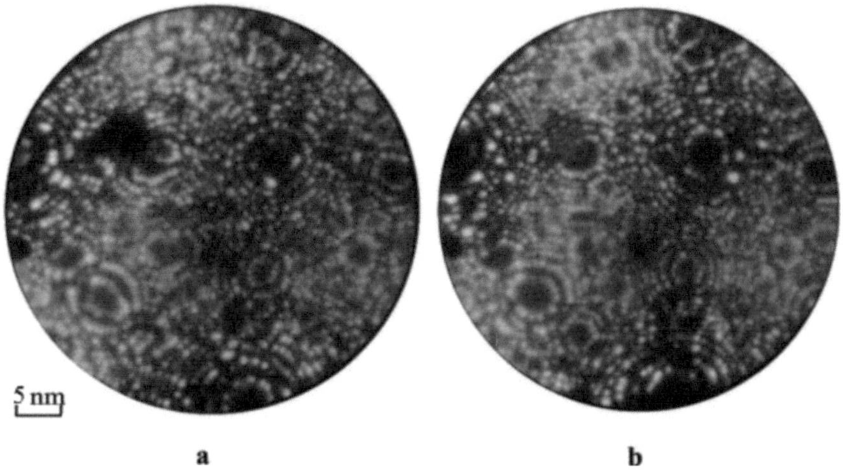

Figure 7.
Ion images of Pt surface after Ar$^+$ irradiation (E = 30 keV, F = 10^{17} ion/cm^2): (a) typical neon contrast of nanopores (shown by arrow); (b) contrast of dislocation loop after field evaporation of 8 atomic face layers (001) (~1.6 mm) (shown by arrow).

of atomic layers, the cross section of the vacancy cluster was fixed. The "exit" of nanopores from a material at the removal of atomic layers, as a rule, came to an end with contrast of a dislocation loop (**Figure 7b**).

The sizes of nanopores were estimated from the ion contrast both using the cross-sectional area of the sample surface defect and the depth of its occurrence and by counting the evaporated atomic layers from the beginning of the nanopore contrast to the complete disappearance of such a contrast. The measurements showed that the nanopores had both spherical and cylindrical shape. According to the authors, the sizes of nanopores in diameter ranged from 1 to 5 nm, in depth—from 1 to 9 nm. Analysis of the structure in the volume up to a depth of 60 nm from the irradiated surface revealed the features of the formation of nanopores. The concentration of nanopores and their distribution in the near-surface volume of irradiated material were established. It was determined that up to 40% of the nanopores are in the near-surface layer with a thickness of 10 nm; in the future the volume fraction of micropores decreased by logarithmic dependence (**Figure 8**). It is known from [15] that argon ions at the used beam energies have a projective range in platinum of no more than 10 nm.

Since nanopores were observed in the metal to a depth of 60 nm, it is obvious that the pore swelling in the material was not due to the presence of implanted argon ions in the defects.Detailed analysis of experimental data allowed assuming that ion implantation resulted in continuous appearance and migration of implanted atoms and vacancies to drains and coalescence of individual vacancies into vacancy nanoclusters (nanopores).

As a result, the formation of nanopores both on the surface and in the near-surface volume of metals in the nanometer range from the irradiated surface was experimentally studied.

The conditions of the threshold exposure at which begins the formation of the nanopore for pure metals are the following: at fluences of 10^{17}–10^{18} ion/cm^2 in the energy range of 20–30 keV, starting from j = 200 μA/cm^2. The distribution and bulk fraction of nanopores in the subsurface material bulk were studied. As a result, it was found that up to 40% of the pores were concentrated in the subsurface layer with a thickness of 10 nm. Later, the bulk fraction of micropores decreased logarithmically. The obtained results can be used for prediction of radiation stability of materials based on fcc metals.

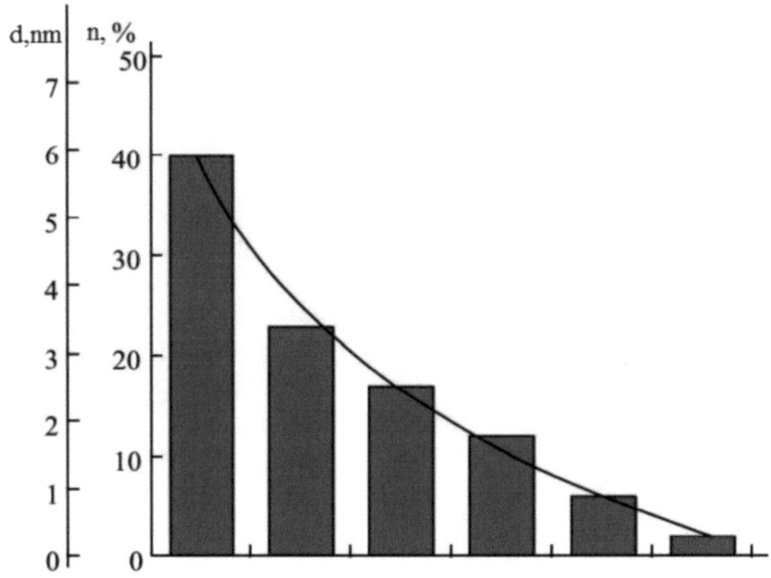

Figure 8.
Concentration of nanopores in platinum irradiated by Ar^+ ions ($F = 10^{18}$ ion/cm^2).

3.3 Experimental modeling of fast neutron beam impact on Pt using Ar^+ beams

Objective: to establish the adequacy of the influence of different types of radiation on the same material (Pt). The comparison of the formed radiation damages of the same type is carried out. The structure of radiation defects on the atomic-pure platinum surface, initiated by neutron (E > 0.1 MeV) and ion interaction (E = 30 keV), was studied by field ion microscopy (FIM).

A large number of radiation defects of the crystal lattice [16] were found in irradiated to fluence 6.7×10^{21} m^{-2} (E > 0.1 MeV) Pt. Among them are single point defects, individual vacancies, displaced interstitial atoms, and vacancy clusters.

Figure 9 shows the ionic contrast of the typical real spatial distribution of radiation damages of the crystal lattice Pt after interaction with neutron radiation in the process of controlled removal of surface atoms is given. Violations of the ring pattern of ion contrast were recorded on the surface of the irradiated platinum. It is for violations in an annular pattern of ionic contrast that is fixed by the defectiveness of the crystal structure. According to the known ion contrast of defects [14], one or another type of radiation defects arising in the material is identified. The change in the ion contrast of the irradiated platinum surface compared to the contrast of the non-irradiated Pt surface is due to radiation damage. This is the result of the interaction of neutrons with the atoms of the crystal lattice. In the process of controlled removal of platinum surface atoms, the structure of defects in the volume was analyzed. The radiation damages detected from the ion contrast were either single point defects (vacancies, interstitial atoms) or small vacancy complexes with dimensions commensurate with the interatomic distances.

Depleted zones (a region with a locally increased concentration of vacancies) with a belt of "interstitial atoms" were found at $F = 3.5 \times 10^{22}$ m^{-2} [16]. This result confirms the hypothesis [17]. According to this hypothesis, the cascade in the metal develops so that a large number of atoms are carried out from its central part (the most perturbed region) by means of chains of substitutions. According to the authors, the concentration of interstitial atoms was 1.5%, and the average concentration of vacancies in the depleted zones was 9%.

The spatial geometry of depleted zones in $F = 3.5 \times 10^{22}$ m^{-2} platinum irradiated to the fluence of intermediate and fast (E > 0.1 MeV) neutrons is analyzed

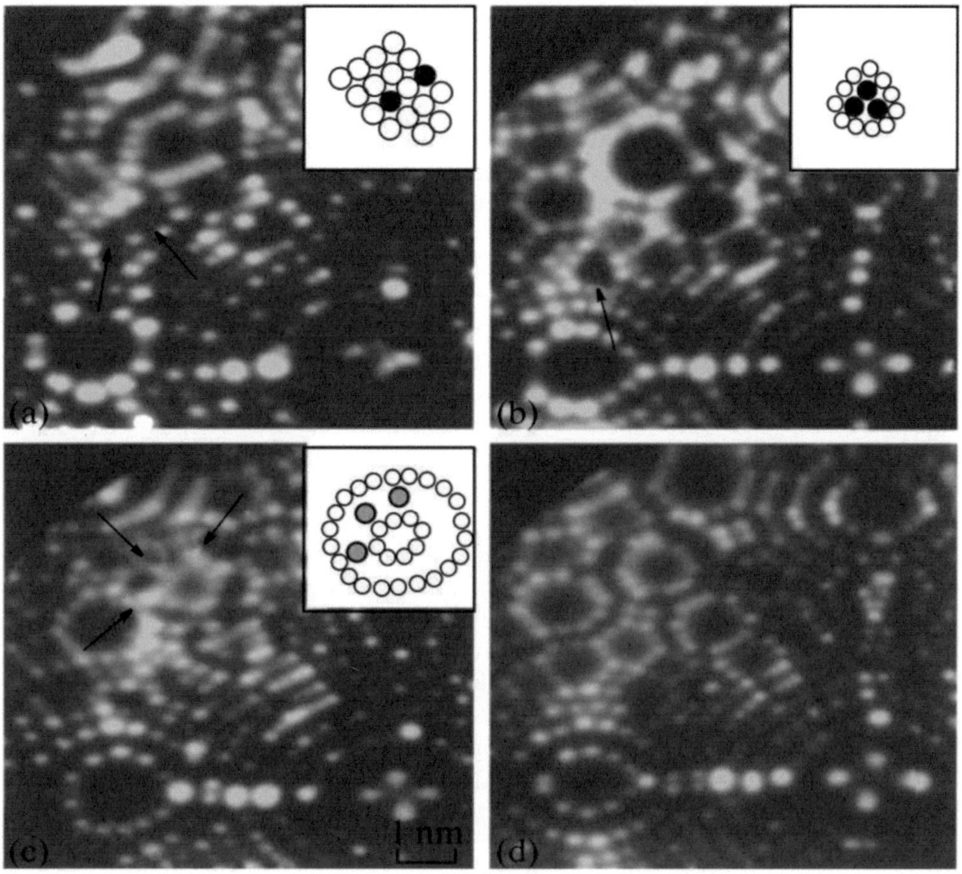

Figure 9.
Contrast of the same surface area irradiated with F = 6.7 × 10²¹ m⁻² Pt. the defects registered in the volume are shown by arrows: (a) single vacancies, (b) tetrahedral vacancy cluster, (c) interstitial atoms, (d) the same surface area without defects. In the upper right corner of the microimages are diagrams of these defects: ○, the image of the atom; •, the image of the vacancy; ●, the image of the interstitial atom.

in [16]. The shape of the depleted zones was determined in the regime of controlled removal of atomic layers to determine the characteristic anisotropy. As a result of the analysis of the ion contrast, no defects of the anisotropy of the depleted zone shape were found. The authors believe that the configuration of the zones does not correspond to any simple geometric figure.

Statistics of a large number of ion images of the surface of irradiated platinum allowed to determine the size of individual depletion zones. From these data, the average diameter of the radiation cluster is 3.2 nm.

As a result of neutron irradiation of Pt up to 6.7×10^{21} m^{-2} ($E > 0.1$ MeV), the structural state of a pure metal somewhat changed. This is evidenced by the microimages corresponding to an atomically clean platinum surface after irradiation. Defective surface regions were observed on the micrographs where the interaction with neutrons led to the displacement of atoms from regular crystal lattice sites. As a rule, radiation damages represented individual vacancies, small vacancy clusters displaced from equilibrium positions (crystal lattice sites), and individual atoms and regions of displaced atoms. Therefore, we can conclude that the effect of the aforementioned neutron fluence in platinum creates separate, nonoverlapping atomic collision cascades.

The preliminary ion images of the field emitters obtained before the ion irradiation showed an almost perfect annular contrast for single crystals of pure metal, indicating the absence of structural defects [4].

The ion image of the surface of platinum irradiated with Ar^+ ions with E = 30 keV; F = 10^{16} ion/cm^2 is shown in **Figure 10**. Ion contrast showed disturbances in the ring pattern of the crystal planes. Such disturbances in the ring pattern of ion contrast are defined as defects in the crystal structure occurring in the material after radiation exposure. Changes in the ion contrast of the irradiated platinum compared to the contrast of the certified Pt were recorded in a layer 1.5-nm thick from the irradiated surface. The ion contrast shows the appearance of a block nanoscale structure in the near-surface volume of the material [4].

Ion contrast of surface radiation damage after ion irradiation (**Figure 10**) differs from the contrast of radiation damage after the interaction of fast neutrons with the material (**Figure 9**). In the analysis of the structure of the near-surface volume Pt after ion irradiation by successive removal of atomic layers at a depth of 2 nm from the surface was identical to the radiation damage in **Figure 9**. **Figure 11** shows the ion contrast of the detected radiation damage.Hence, irradiation by fast neutrons (E > 0.1 MeV, F = 6.7 × 10^{21} m^{-2}) leads to the formation of the same amount of radiation damages which is observed at a depth of 2 nm after the Ar^+ irradiation (E = 30 keV, F = 10^{16} ions/cm^2).

It follows that all radiation damages in the volume of the material caused by neutron irradiation are lattice defects that are determined by the neutron fluence and energy. These defects do not depend on the depth of irradiation. The type of radiation damage caused by the interaction of charged ion beams with the substance is determined in addition to the irradiation parameters by the distance from the irradiated surface. It follows from the results that the process of transition of one type of radiation damage to another depends on the depth of irradiation of the material. To simulate the analog effects of fast neutrons with ion irradiation by comparing the number of defective structures is possible only at a certain depth which is irradiated with beams of charged particles of matter. Therefore, to achieve this task, it was necessary to detect the identity of the defective structure in different types of radiation as well as for ion irradiation at the appropriate depth of the near-surface volume of the material. Thus, we have established that the impact of fast neutrons interacting with matter corresponds to the interaction of positively charged Ar^+ ions at a certain depth from the irradiated surface. Therefore we can simulate the effect of neutron irradiation by using ion beams based on the arising particular radiation damage structure in the material.

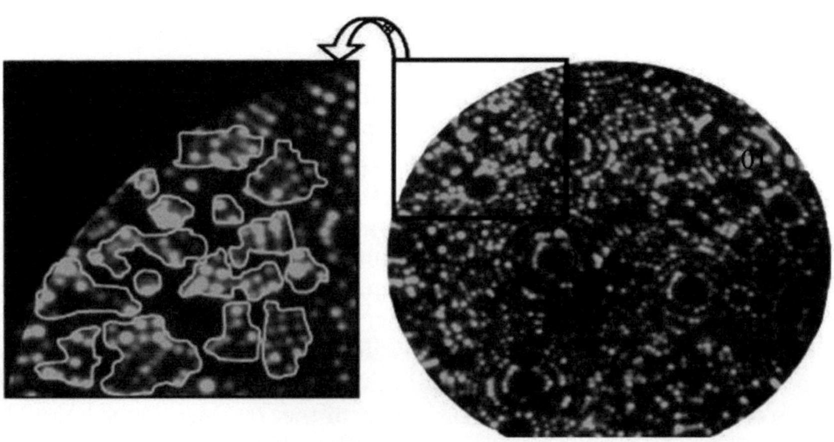

Figure 10.
Micrographs of the Pt surface region after irradiation by Ar^+ ions with F = 10^{16} ions/cm^2 (T = 70°C) obtained using neon as the imaging gas. Nanoblocks are indicated.

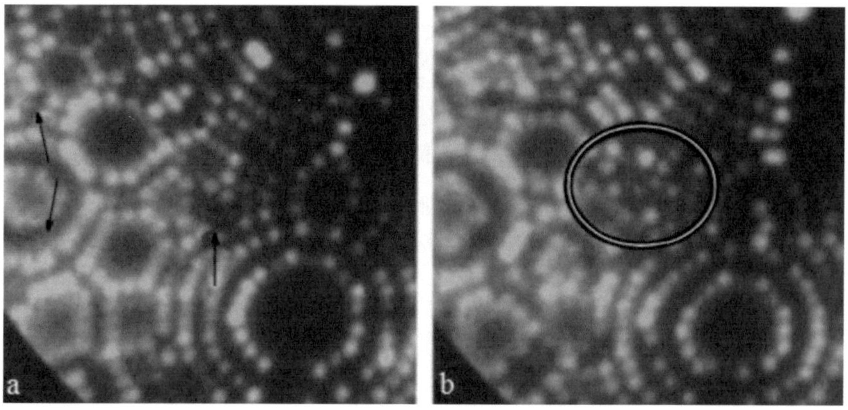

Figure 11.
*Same as in **Figure 10**: (a) at a depth of 2 nm from the irradiated surface (arrows indicate isolated vacancies and interstitial atoms); (b) at a depth of 1.5 nm from the surface (the depleted zone is shown).*

3.4 Cascades of atomic displacements in metals and alloys after various types of irradiation

The atomic structure of defects in an atomically ordered Cu_3Au alloy was studied after irradiation with beams of charged particles. The tip samples were irradiated perpendicular to the needle axis with E = 40 keV: at ion current density j = 10^{-3} A/cm^2 and pulse duration $\tau = 10^{-3}$ s. The radiation dose was taken D ~ 6 × 10^{16} ion/m^2 to provide (on average) one ion per element with an area of 4 × 4 nm. Thus, the formation of separate single stages of displacement in the volume of the material was achieved. The analysis of ion micrographs of the surface showed the formation of such radiation defects as disordered zones and segregation of copper atoms [18]. The average size of the disordered zones was estimated from the analysis of the structure in the amount of between four samples which is 4 × 4 × 1.5 nm. Disordered zones were found in the form of a violation of the ion contrast of the ring pattern of the image of the surface of an atomically ordered alloy. The ionic contrast of Cu_3Au alloy in the ordered state is similar to the ionic contrast of the pure metal surface, because this contrast is formed only by gold atoms. The contrast of radiation defects was observed in the process of controlled removal of surface atoms to the depth of disordered zones. The experimentally measured average size of the zones coincided with the calculated one in order of magnitude. The calculated diameter of the displacement cascade was 5–11 nm. The latter was calculated as the mean free run of PVA in the approximation of the spherical shape of the cascade region. It can be assumed that the real size of the cascade regions (taking into account the mean free path of atoms knocked out of the cascade nucleus and occupying intervals that cannot always be reliably identified by FIM), apparently, is slightly larger than the experimentally measured size.

During the analysis of the ion contrast of the defective regions, only segregation of copper atoms was found. The contrast from the segregation of copper atoms is observed as a dark region due to the absence of images from copper atoms in the ionic images of ordered Cu_3Au. It was found that the segregation of copper atoms is three-dimensional and usually contains 200–500 atoms. The analysis of the ionic contrast of the dark area boundaries shows that the possible number of vacancies in the segregation is insignificant.As a result of neutron bombardment of platinum to a fluence of 6.7 × 10^{21} m^{-2} (E > 0.1 MeV), the structural state of pure metal has changed; this follows from **Figure 12**, where the atomically pure surface of irradiated platinum is recorded. Defective regions of the surface marked by ovals on

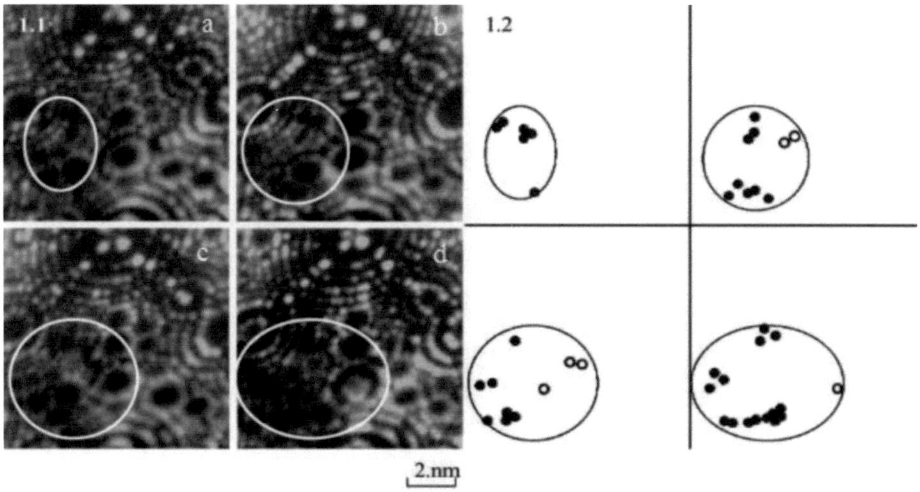

Figure 12.
Regions of sequential field ion images of the platinum surface bombarded by neutrons (I) and corresponding scheme of spatial distribution of defects (II). The depleted zone formed as a result of sequential controllable removal of atomic layers (a–c): (a) initial distribution; (b) differs from (a) by 1 ac; (c) differs from (b) by 2 ac; (d) differs from (c) by 2 ac: •, vacancy; ○, interstitial atom.

microphotographs are the result of the interaction of neutrons with atoms in the crystal lattice. As a rule, the defects detected in the course of piecewise removal of surface platinum atoms by an electric field have the form of individual vacancies, small vacancies complexes, and individual atoms displaced from equilibrium positions (crystal lattice sites) [16].

Analysis of the structural state of pure Pt after neutron bombardment with a fluence increasing to 6.7×10^{22} m^{-2} ($E > 0.1$ MeV) revealed the presence of a large number of crystal lattice defects, which is visually demonstrated in **Figure 12**. As a result of controllable removal of surface atoms of platinum by an electric field, radiative clusters in the form of 3D "loose" regions containing a large number of point defects (individual vacancies, interstitial atoms, and small vacancies complexes) were detected apart from single point defects and their aggregates. **Figure 12-I** shows the ion contrast of a typical actual spatial distribution of distortions in the Pt crystal lattice after the interaction with fast neutron beams in the course of controllable removal of surface atoms. The corresponding schematic chart is shown in **Figure 12-II**. Analysis of the ion contrast of the defective region under investigation showed that the latter is a depleted zone (region with a locally elevated concentration of vacancies) with the "belt" of interstitial atoms. This observation confirms the hypothesis [17] according to which a cascade in the metal evolves so that a large number of atoms are carried away from its central part (the most perturbed region) by chains of substitutions. According to our estimates, the average concentration of vacancies in depleted zones amounts to 9%, while the concentration of interstitial atoms is 1.5%.

We have made an attempt at clarifying the spatial geometry of depleted zones in platinum bombarded by fast neutrons to a fluence of 6.7×10^{22} m^{-2} ($E > 0.1$ MeV). The shape of depleted zones was analyzed in the standard regime by controllable evaporation of atomic layers by an electric field for determining the characteristic anisotropy. As a result of quantitative analysis of the geometry of defective regions, no anisotropy in the shape of depleted zones has been revealed. Experimental data show that the configuration of these zones does not correspond to any simple geometrical figure because vacancies constituting these zones are arranged extremely irregularly.

Analysis of a large number of microphotographs of the irradiated platinum surface has made it possible to measure the sizes of distorted regions (determined either by calculating the radii of curvature or the tip of the emitter and corresponding linear distances or by calculating the number of atomic layers evaporated by the field), to construct the histogram of their size distribution (**Figure 13**), and to determine the average diameter of a radiation cluster, which amounted to 3.8 nm.

Apart from dimensional characteristics of clusters, we estimated their concentration. The experimentally determined number density of such radiation-induced distortions in the bulk was 4×10^{24} m^{-3}.

After the bombardment of platinum to a fast neutron fluence of 6.7×10^{21} m^{-2} ($E > 0.1$ MeV), a defective structure formed, which was characterized by an elevated concentration of single point defects and their complexes with sizes comparable to the atomic spacing. As the fluence increased to 6.7×10^{22} m^{-2}, radiation clusters (depleted zones with a belt of interstitial atoms) appeared in the irradiated platinum; the average size of these clusters was 3.8 nm.

As a result of a direct study of the atomic structure of radiation defects in the near-surface volumes of metal materials, quantitative estimates of the size, shape and volume fraction of individual cascades of atomic displacements are obtained. At the atomic-spatial level, the analysis of radiation defects of various types after neutron and ion irradiation was carried out. The formation of such defects is due to the development of single cascades of atomic displacements and the ongoing relaxation processes of radiation-stimulated diffusion and segregation.

3.5 Analysis of the mechanisms of radiation-induced effect of nanostructuring

Prior to irradiation of pure platinum with beams of charged Ar^+ particles, the tip samples were certified in FIM, that is, they had an atomically smooth surface. Ionic contrast from these samples had virtually no violations of ring patterns (**Figure 14a**).

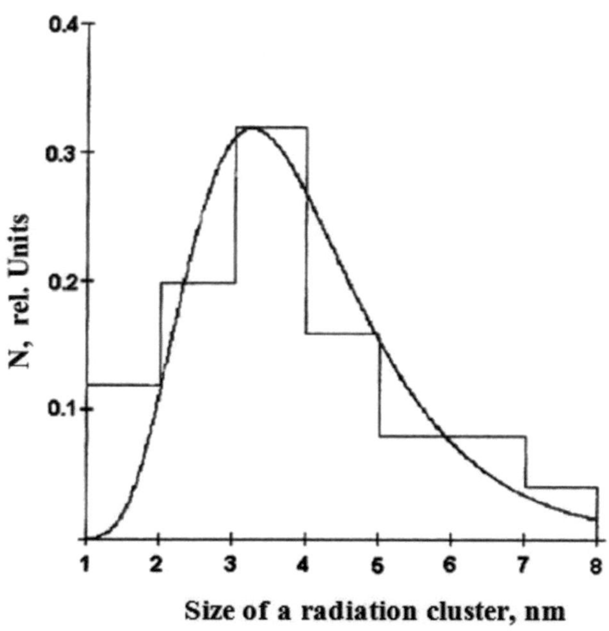

Figure 13.
Size distribution of radiation clusters in platinum bombarded by neutrons (6.7×10^{22} m^{-2}).

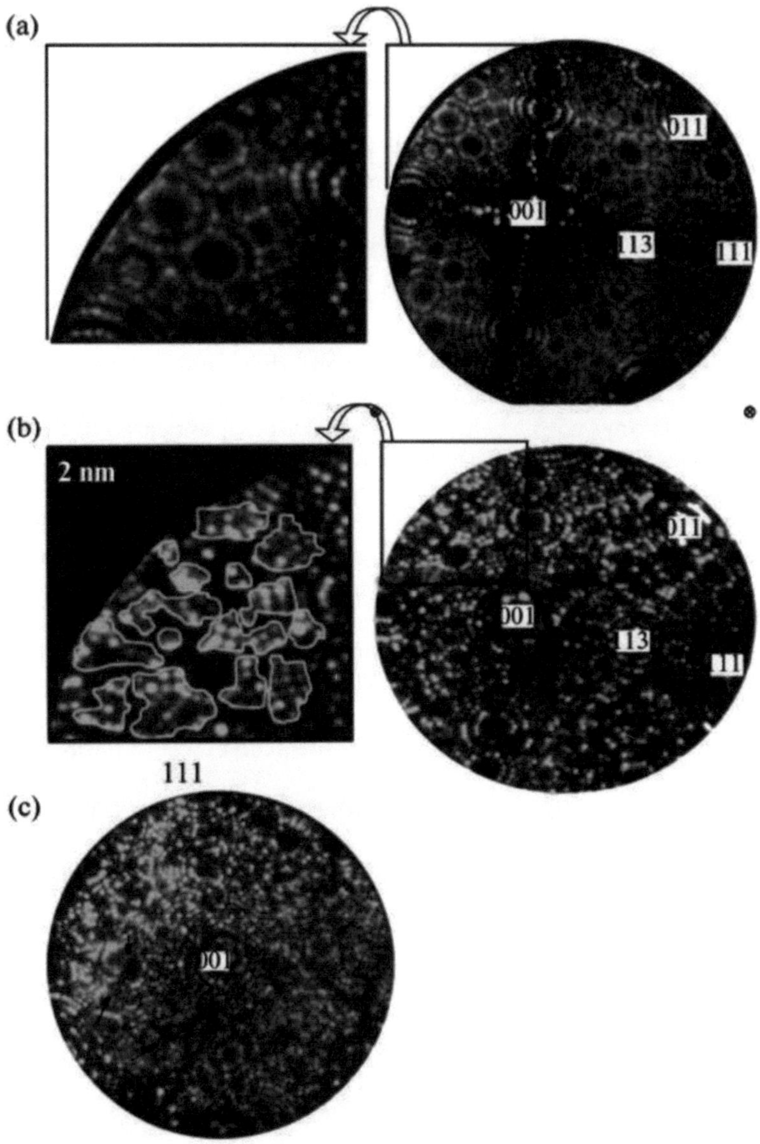

Figure 14.
Neon FIM mages of Pt points (a) in the initial (attested) state, (b) upon irradiation with 30 keV Ar^+ ions to $F = 10^{16}$ ion/cm^2 (T = 70°C), and (c) upon irradiation with 30 keV Ar^+ ions to $F = 10^{17}$ ion/cm^2 (T = 200°C). Arrows indicate the typical ion contrast of nanoparticle boundaries and stacking faults.

After Pt irradiation with E = 30 keV, $F = 10^{16}$ ion/cm^2, the ring contrast pattern changed significantly (**Figure 14b**). On the image of the surface, there were violations of the ring pattern of ion contrast, which are proof of defects in the volume of the material. Changes in ionic contrast (**Figure 14a** and **b**) metal in the initial (certified) and irradiated state register a block nanoscale structure in the near-surface volume of the material. This block structure was observed in a layer 1.5 nm thick from the irradiated surface.

In [4], a quantitative analysis of the nanoblock size distribution in the irradiated to fluence $F = 10^{16}$ ion/cm^2 near-surface volume ($V \approx 250$ nm^3) of platinum was performed. The linear method of Rosival, based on the Cavalier-Acker principle, was chosen for the determination of nanoblock volume fractions of different sizes.

Analysis of ionic contrast image of atoms in nanobiotech (**Figure 14b**) showed that atoms are located almost at the nodes of the crystal lattice. The blocks themselves were disoriented relative to each other. It is obvious that the target is a single crystal of platinum (the object of study in the microscope, having a radius of curvature of 30–50 nm, is almost always a single crystal). Therefore, the authors suggest that the main mechanism of formation of the nanostructured state at 10^{16} ion/cm^2 fluence is associated with the phenomenon of channeling [19].

The effect of the formation of a block nanocrystalline structure varies in the size range with increasing fluence to F = 10^{17} ions/cm^2 (**Figure 14c**). The size of the blocks increases to 1–5 nm and is observed in the near-surface layer with a thickness of at least 20 nm from the irradiated surface.

From the experimental data obtained, the transverse and longitudinal dimensions of nanocrystalline blocks (**Figure 15**) and the width of the boundary area between the nanoblocks were determined. The width of the boundary region, according to our estimates, varied from 0.4 to 0.8 nm at various sites of nanoblock boundaries in irradiated platinum.

Ionic contrast observed on the irradiated platinum surface shows typical contrast for grain boundaries and stacking faults [14]. This contrast is characteristic and is observed in the images of almost all faces of nanocrystals (**Figure 15**).

Based on this, it follows that at F = 10^{17} ions/cm^2, the mechanism of formation of the nanoblock structure in the body of the material changes.

It was previously shown [20, 21] that similar nanocrystalline block structures can also be formed as a result of intense plastic deformation. Based on these data, it can be assumed that the nanoblock structure found in this work is (with an increase in the fluence to F = 10^{17} ion/cm^2) the result of deformation processes occurring in the material during ion irradiation and in the subsequent period of time after irradiation.

Experimental results obtained in [22] can serve as confirmation of such interpretation of the phenomenon of formation of the nanostructured near-surface layer. The authors [22] claim that point defects, dislocation loops, and dislocations

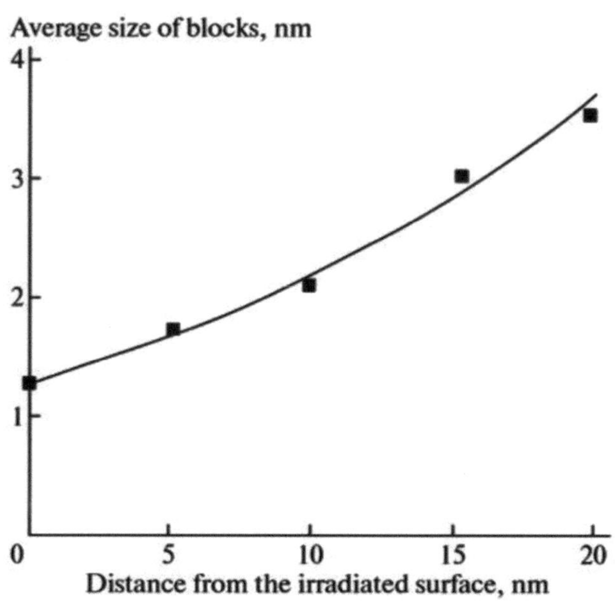

Figure 15.
Variation of the average block size in depth of the transverse cross section of a sample irradiation with 30 keV Ar^+ ions to F = 10^{17} ion/cm^2.

are generated in the surface layer doped during ion implantation. Dislocation substructures (DSS) can also be formed in the near-surface zone. According to the classification of DSS [23], nanostructural states may also occur as a result of irradiation during complex dislocation evolutions.

Therefore, one of the proposed mechanisms of the nanostructural states in the near-surface volume of pure metals irradiated to F = 10^{17} ion/cm^2 can be the deformation model.

As a result, the effect of ion implantation on the crystal structure of platinum was studied experimentally, on the atomic-spatial scale, when the irradiation parameters (energy, ion current density, and radiation dose) were changed by accelerated beams of charged argon ions. The mechanism of occurrence of nanostructured states in the near-surface volume of metals in the nanometer range from the irradiated surface is analyzed from the data obtained in FIM.

It is shown that the modification of the crystal lattice in the near-surface volume of irradiated platinum depends on the modes of irradiation by charged, accelerated to 30 keV, Ar^+ ions. At fluence F = 10^{16} ion/cm^2, the phenomenon of nanostructurization of the near-surface volume can most likely be explained by the channeling effect.

An increase in the fluence by an order of magnitude leads to a change in the mechanisms of nanostructures in the irradiated volume. The observed deformation character of the ion contrast of the metal surface, registered in the process of controlled sequential removal of atomic layers, and the subsequent analysis of the state of the near-surface volume from the experimental data, allows us to judge the prevalence of the deformation model explaining the mechanism of nanostructurization.

Thus, modification in the nanometer range of the near-surface volume of irradiated metals occurs due to several mechanisms. In particular, for fluence F = 10^{16} ion/cm^2 (E = 30 keV), the main contribution is made by the channeling effect. Increasing the fluence by an order of magnitude leads to the predominance of the deformation mechanism in the formation of nanostructured States in the near-surface volume of the metal.

It can be assumed that the formation of nanostructured states in the near-surface volume of the material as a result of implantation will lead to a significant increase in the physical and mechanical properties of the irradiated substances.

4. Summary

Diagnostics of the irradiated crystal structure of metal materials was carried out by field ion microscopy.

Determine the modes of radiation exposure for creating amorphized states in subsurface volume of platinum. It is shown that the radiation effect on pure metals with E = 30 keV when the fluence of charged argon ion beams changes by two orders of magnitude (from 10^{16} to 10^{18} ion/cm^2) significantly affects the kinetics of defect formation in the near-surface volume of irradiated materials. It was found that the phenomenon of metal amorphization in the near-surface volume occurs up to 12 nm in the depth of the sample with an increase in the fluence to 10^{18} ion/cm^2, at the above irradiation energies.

The threshold of formation of nanopores in the irradiated platinum was established. The threshold in the irradiated platinum corresponds to the fluence F = 10^{17} ion/cm^2. The size of the nanopores was determined: in the transverse, from 1 to 5 nm, and longitudinal (in the depth of the target), from 1 to 9 nm. It was found that up to 40% of the nanopores are concentrated in the near-surface layer with a thickness of 10 nm.

It is shown that the interaction of fast neutrons (E > 0.1 MeV) $F = 6.7 \times 10^{21}\ m^{-2}$, $F = 3.5 \times 10^{22}\ m^{-2}$ with the substance leads to the formation in the platinum volume of the same radiation damage that occurs after ion irradiation with charged ion beams Ar^{+} C E = 30 keV, $F = 10^{16}$ ion/cm^2, and is observed in Pt at a depth of about 1.5–2 nm (respectively, $F = 6.7 \times 10^{21}\ m^{-2}$ and $F = 3.5 \times 10^{22}\ m^{-2}$) from the irradiated surface. As a result, the effect of neutron irradiation by charged ion beams on the basis of the appearance of a certain structure of radiation damage in the material is experimentally simulated.

The results of atomic-spatial study of radiation defect formation in near-surface volumes of materials initiated by neutron irradiation, Pt: E > 0.1 MeV, and ion implantation, Cu_3Au: E = 40 keV, $F = 10^{16}$ ion/m^2, $j = 10^{-3}$ A/cm^2, are presented. Quantitative estimates of the size, shape, and volume fraction of atomic displacement cascades formed under different types of irradiation in the near-surface layers of materials are presented. It is shown that the average size of radiation clusters after irradiation of platinum to the fast neutron fluence of $6.7 \times 10^{22}\ m^{-2}$ (E > 0.1 MeV) was about 3.8 nm. Experimentally established average size of radiation clusters (disordered zones) of the alloy after ion irradiation amounted to the value of $4 \times 4 \times 1.5$ nm^3.

The mechanisms of radiation-induced effect of nanostructurization of near-surface volumes of metals are analyzed. It is assumed that the modification of the surface volume of metals in the nanometer range when interacting with charged particle beams Ar^{+} occurs due to several mechanisms. In particular, for fluence $F = 10^{16}$ ion/cm^2 (E = 30 keV), the main contribution is made by the channeling effect. Increasing the fluence by an order of magnitude leads to the predominance of the deformation mechanism in the formation of nanostructured states in the near-surface volume of the metal.

Author details

Vladimir Alexandrovich Ivchenko[1,2]

1 Institute of Electrophysics, Ural Branch, Russian Academy of Sciences, Yekaterinburg, Russia

2 Yeltsin Ural Federal University, Yekaterinburg, Russia

*Address all correspondence to: ivchenko2008@mail.ru

References

[1] Guseva MI. Ion Implantation in Semiconductor Materials, Developments in Science and Technology. Series: Physical Principles of Laser Beam Technology. Vol. 5. Moscow: VINITI; 1989. [in Russian]

[2] Ovchinnikov VV. Izvestiya Akademii Nauk SSSR. Mössbauer spectroscopy of ion-alloyed metals and alloys. Metally. 1996;**6**:104

[3] Poate JM, Foti G, Jackobson DC, editors. Surface Modification and Alloying by Laser, Ion, and Electron Beams. New York, London/Moscow: Plenum Press/Mashinostroenie; 1983/1987

[4] Ivchenko VA, Medvedeva EV. Modification of nano-structured states in ion_implanted platinum. Izvestiya Rossiiskoi Akademii Nauk. Seriya Fizicheskaya. 2010;**74**(2):237

[5] Ivchenko VA, Syutkin NN, Kuznetsova LY. Amorphization of the subsurface regions in ion-implanted alloys. Technical Physics Letters. 2000;**26**:541

[6] Ivchenko VA, Syutkin NN, Bunkin AY. FIM investigation of ion-implanted Cu_3Au alloy. Journal de Physique. 1988;**49–C6**:379

[7] Ivchenko VA, Ovchinnikov VV, Goloborodsky BY, Sytkin NN. FIM of vacancy clusters in the subsurface volume of the ion-implanted Pd(CuAg) alloy. Surface Science. 1997;**384**:46

[8] Ivchenko VA, Syutkin NN. Field ion microscopy of deformation effects in the subsurface volume of ion-implanted metals (Ir). Pis'ma v Zhurnal Tekhnicheskoi Fiziki. 1999;**25**(6):410

[9] Ivchenko VA, Medvedeva EV. Modification of nanostructured states in ion-implanted platinum. Bulletin of the Russian Academy of Sciences: Physics. 2010;**74**:217

[10] Ivchenko VA, Medvedeva EV. Field ion microscopy of nanostructures in HCC metals. Perspektivnye Materialy. 2009;**7**(Spec. Issue):119

[11] Myuller EV. Field ionization and field ion microscopy. Physics-Uspekhi. 1962;**77**:481-552

[12] Gomer R. Kataliz i Elektronnye Yavleniya [Catalysis and Electronic Phenomena]. Moscow: Inostrannaya Liter Atura; 1958

[13] Ivchenko VA. Field ion microscopy of nanomaterials after intensive external influences. International Journal of Nanomaterials and Nanostructures. 2012;**XXI**(1):42

[14] Boowkett KM, Smith DA. In: Amelinckx S, Gevers R, Nihoul J, editors. Defects in Crystalline Solids. Vol. 2. Amsterdam/London: North Holland Publishing Company; 1970. p. 257

[15] Burenkov AF, Komarov FF, Kumakhov MA, Temkin MM. Spatial Distributions of Energy Evolved in a Cascade of Atomic Collisions in Solids. Moscow: Energoat Omizdat; 1985. [in Russian]

[16] Ivchenko VA, Medvedeva EV, Ovchinnikov VV. Journal of Surface Investigation: X-ray, Synchrotron and Neutron Techniques. 2009;**3**:592

[17] Orlov AN, Trushin YuV. Theory of radiation cascades in crystals. Questions of atomic science and technology. Series: Radiation Damage Physics and Radiation Material Science. 1985;**2**(35):14

[18] Bunkin Y, Ivchenko VA, Kuznetsova LY, et al. Field-ion microscopy of defects

in regions of displacement cascades in alloy Cu_3Au. The Physics of Metals and Metallography. 1990;**70**:104

[19] Nastasi M, Mayer JW, Hirvonen JK. Ion Solid Interactions: Fundamentals and Applications, Cambridge Solid State Science Series. Vol. 27. Cambridge: Cambridge University Press; 1996

[20] Ivchenko VA, Syutkin NN. Effect of low-energy (20-40 keV) ion implantation on phase transformations in the subsurface volume of alloys. Applied Surface Science. 1995;**87-88**(C):257

[21] Ivchenko VA, Efros BM, Popova EV, Efros NB, Loladze LV. Field ion microscopy of metals under intensive external action. Fizika i Tekhnika Vysokikh Davlenii. 2003;**13**(3):109

[22] Didenko AN, Sharkeev YR, Kozlov EV, Ryabchikov AI. Long-Range Interaction Effects in Ion-Implanted Metal Materials. Tomsk: Izdat. NTL; 2004. [in Russian]

[23] Koneva NA, Kozlov EV, Trishkina LI. Classification of dislocation substructures. Metallofizika. 1991;**12**(1):49

Chapter 8

Chitosan Formulations: Chemistry, Characteristics and Contextual Adsorption in Unambiguous Modernization of S&T

Rajendra Sukhadeorao Dongre

Abstract

Since long scientists explored natural/bio-polymers to explicit their innate features to develop certain novel utilities in modernization of prevalent Science & Technology. Consequently biotope derived polysaccharide embrace huge prospective desired functions. Amid, chitosan, the second most ubiquitous polymer after cellulose exists as a β-(1–4)-linked d-glucosamine/N-acetyl-d-glucosamine randomly distributed linear polycationic yield from partial deacetylation of chitin polysaccharide. Chitin's complexity limits its extraction/insolubility in aqueous solution, thus less studied/research until 1980s. As major polysaccharides are either neutral/negatively charged in an acidic environment, instead chitosan is cationic, eventually forms electrostatic complexes/multilayer structures/composites with anionic synthetic dopants/natural polymers. Chitosan own biocompatibility, non-toxicity, low allergy and biodegradability allow utility as in water treatment, wound-healing, pharmaceutical excipient/drug carrier, obesity treatment and scaffold for tissue engineering. It is reflected in the increasing number of related publications throughout in biomedical, environmental and industrials applications. Feeble chitosan solubility limits their applications, yet benign synthetic techniques viz.; sol-gel, encapsulation, chemical grafting are employed to yield composites/hydrogels/films/granules which generates new functionality, besides enhanced biocompatibility and biodegradability. This chapter presents the R&D, trends and the latest prospects involved in advance synthesis of chitosan supported composites/hydrogels/films/granules/sheets with special highlighted pharmaceutical/biomedical and environmental applications.

Keywords: chitin, chitosan, pharmaceutics, biomedical, environment, sol-gel, composite, hydrogel, formulations

1. Introduction

Bio-polymers yield via flora and fauna; plants, fungi and many other natural origins constantly attentive the worldwide researchers by virtue of endurance for atmosphere and our life [1]. Among bio-molecules polysaccharide like chitin/

chitosan which carried skeletal architecture in many animals besides feedstock used in rational designing of smart materials consequently [1, 2]. S&T vitally explored chitin/chitosan for comprehensive growth and economic progression in assorted fields including clinical, medical, pharmaceutics and environment along with fulfilling sophisticated nanotechnology requirements [3]. Today scientific modernism and industry inventively carry out R&D in pharmaceutics, environment, and bio-technology which fortify livelihood and offers accessible facilities via copious trustworthy merchandise [1–4]. For this purpose, chitin/chitosan matrix is fore-mostly investigated to derive innovative formulations owing innate exclusive, multi-functional and particular variable characteristics devoid in customary stuff and counter parts [1–6].

Chitin own β-[1,4]-2-acetamido-2-deoxy-D-glucose/N-acetylglucosamine monomeric unit as linked via glycosidic bonds which occurs as the second copious natural polymers after cellulose to cater numerous remarkable prospective needs in prevalent modernization [1]. Chitin is regularly produced all over the world with capacity of 1500 tons/annum. Chitin/chitosan is frequently utilized in biochemistry, microbiology, chemistry, polymer engineering, pharmacy, medicines and material sciences [4]. Flexible raw chitin undergoes alkaline deacetylation to derive chitosan matrix own further facile molding via assorted biological/physicochemical amendments to yield better capable composites, hybrids and blends exceeded over counterpart cellulose. Chitin/chitosan owe elite characters viz.; highly flexible, bio-compatible, bio-degradable and non-toxic. Nano-technology signified chitin chemistry in S&T via preparation of various innovative, creative and widely usable matrix formulations from lithe chitosan [7, 8]. Model chitin/chitosan matrix can be improved which offers requisite beneficial applications in recent modernization [1–9]. Thus, chitosan is extraordinarily impressive to derive easy formulations accredited to innate proactive —NH_2/—OH functionalities as executed via varied physic-chemical chemical alterations viz.; —NH_2/—OH acylation or alkylation and primary amine N-quaternization besides C-6 carboxylation [1]. Extensive adaptations in chitosan imparts inclusive cationic characters due to protonation of primary —NH_2 to ammonium ion —NH_3^+ which resulted acid to alkaline pH dependency and extra solubility [2–9]. Liberal breakthrough are offered in many fields including nanoscience, biotechnology, pharmaceuticals and tissue engineering scaffolds procured via chitin/chitosan outstanding character viz.; biodegradable, biocompatible, non-toxic and antifungal/microbial immunogenic profiles [9]. Some chitosan derived superior formulated materials in nanoscience/biotechnology along with case study of fluoride mitigation from water are summarized in this chapter.

2. Chitin-chitosan chemistry

2.1 Mucoadhesiveness

Chitosan is mucoadhesive due to inherent cationic nature and hydrophobic interactions that found weaker than anionic polymeric carbomer. Sustainable mucoadhesive character offers high cohesive/adhesive bonds within polymeric matrix as comparatively weak mucus gel layers. The rational chemical, biological or physical treatments on raw chitosan framework gets improved via complexation with multivalent anionic excipients like inorganic/organic ionic drug components. Some strategic alterations in its skeletal are partial due to cationic substructures imparting ionic interactive mucoadhesion. Oral bioavailability involving with such mucoadhesive chitosan particularly not gets achieved if mixed with polyanionic carbomer. However cationic character and ultimately mucoadhesive properties can be enhanced up to 3/4-fold by trimethylation at NH_2 functionality via PEGylated derivatization or immobilization of thiol groups. Chitosan forms disulfide bonding with mucus gel layer glycoproteins yields most mucoadhesiveness [10]. Gelling material: Macromolecular polymer gel called hydrogel can be constructed through the cross-linked polymeric network using hydrophilic monomers by chain/step growth, besides a purposeful cross-linker employed to endorse net-like structure owing void imperfections. Such hydrogel absorbs water via hydrogen bonding, resulted self-healing alike to expand typical firmness hitherto mechanical elasticity. Self-healing spontaneously forms new bonds within a hydrogel matrix during reconstructive covalent dangling surface chaining or via non-covalent hydrogen bonding. Inherently flexible chitosan skeleton have motivated the R&D of self-healing hydrogels are invoked as reconstructive tissue engineering scaffolds besides in passive and preventive utility.

Chitosan hydrogels are resulted for *in-situ* gelling which is properly altered via pH-dependent hydrostability [1–4]. In-situ gelling delivery system derived from chitosan and polyacrylic acid blend yields liquid state formulation at moderate acidic pH which gets transformed into viscous gel at pH 7.5. Chitosan's —OH/NH_2 functionality undergoes cross-linking via disulfide treatment/thiolation found to impart additional *in situ* gelling due to significant viscosity use to access oxygen on nasal/ocular mucus surfaces. Chitosan–thioglycolic acid conjugate cross-linked hydrogels are rationally designed owing 16,500-fold instant viscosity to be utilized for advanced clinical uses [11]. Gene expression material: Chitosan skeleton gets modified to impart gene expression characteristics. Strategic self-branching of chitosans improved its basic gene transfer properties without conciliation its innate safety domain [12]. Trisaccharide moiety with molecular mass of 11–71 kDa are facile to get self-branched onto chitosan framework showed elevated transfection efficiency for gene expression two to five times than that of own linear counterparts. Chitosan/plasmid NP matrix have shown enhanced gene expression levels due to strategic self-branching which found to result in higher stability properties toward nucleases [13]. Reducing conditions of cytoplasm, plasmid gets released in target cells due to framed disulfide bondings which are cleaved *in-situ* and release at the target site. Transfection rate of thiolated chitosan-plasmid NP matrix is fivefold superior to unmodified chitosan/pDNA-NP complex [14]. Advanced gene expression materials owing enhanced cationic characteristics to work as tools for DNA-based drug delivery are developed through assorted physicochemical treatments onto raw chitosan skeleton viz.: trimethylation at primary amino function, cyclodextrin derivatization, and PEG-alkylation [15]. Chitosan based stable complexes of poly-anionic drugs own interfering RNAs/DNAs can perform its controlled/sustainable release. Chitosan being less toxic its higher ratio gets utilized than other cationic counterpart like poly-arginine and poly-lysine, to yield stable complexes with anionic drug moiety. These chitosan complexes protects degradation by DNAses and ultimately imparts high stability and resultant NPs smaller than 100 nm have shown effective positive zeta potential as vital for endocytosis and best for non-viral gene carrier [16].

2.2 Permeation characters

Chitosan own —NH_2/—OH functionality which own pronounced cationic characters and makes it facile for the permeation enhancement due to interactive structural reformation of tight junction-associated proteins [1]. Rationally designed degree of deacetylation and molecular mass of raw chitosan found to control permeation enhancement and toxicity besides fairly more epithelial permeability to the greater extent [1–5]. Chitosan is facile to blend with assorted permeation enhancer doping agents can lead synergistic effective phenomenon resulted 4-fold improved activity. Chitosan-cyclodextrin derived nanoparticles exhibit more permeation enhancement for small peptides as a carrier, nevertheless 30-fold additional permeation enhancement is achieved on certain mucosal membranes via thiol formation/derivatization of chitosan [6]. Many chitosan blends/composites are employed for effective drug delivery of assorted biomaterial/drugs in the treatment of cancer [1], optical [2], and colon diseases [4] as shown in **Table 1**. Based on inherent chitosan's molecular weight of low, medium and higher the *N*-trimethylated with polyethylene glycol (PEG) derivatives are formulated as hydrogel and used in nasal drug delivery [1]. The high/moderate molecular weight chitosan with *N*-trimethylation after blending PEG found to have shorter sol-gel transition span at physiological temperatures and good hydrophilic property besides strong mucoadhesiveness. Such hydrogel formulations of chitosan showed advantageous features like good pharmacokinetic, rheological and mucoadhesiveness, slow/control/sustainable improved drug delivery, fair compatibility encourage stability, more efficacies and low toxicity along with fast sol-gel transition at ambient temperatures [1].

Chitosan-nanostructured frameworks involved rationally designed biological/physicochemical cross-linking within its matrix via alteration at hydroxyl/primary amino functionalities. Cross-link skeleton are reframed as stable droplets via emulsion followed by high-speed stirring random collisions, while precipitation is elicited by coalescence of chitosan droplets with alkali solutions. Ultrafine chitosan NPs

Chitosan formulations	Utility	Applications
Chitosan derived nanogels	For optical pH-sensing analysis	Responsive hybrid nanogels derived from chitosan shown nonreversible pH-sensitivity. Highly stable chitosan-nanogel quantum dots are also used.
Zinc-pectin-chitosan hybrid	Resveratrol drug delivery in colon	1% chitosan formulation with pectin/drug 3:1 ratio at pH 1.5 exhibited best performance of drug release at colon.
Chitosan-zidovudine composite	Stops Zidovudine loss in human plasma, with long life	Extended retention time (shelf life) for composite that gathers in kidney than heart, liver, spleen, lung and brain.
Sodium alginate-chitosan composite	For vaginal delivery of drug	Chitosan-sodium (ratio 1:4, w/w) alginate composite shown controlled release of chlorhexidine digluconate drug.
Cyclosporin A-chitosan hybrid	Extraocular administration	Enhanced therapeutic index of challenge drugs used in extraocular diseases.
Chitosan-based polyelectrolyte coats	Drug delivery on skin	Films/coats viable for notable drug release/permeation through skin.
Chitosan nanospheres loaded by 5-fluorouracil	Delivery of 5-fluorouracil for cancer treatment	These stable nanosized chitosan particles can entrap and deliver drugs in tumor cells.
Chitosan-tripolyphosphate composite own drug	Insulin delivery for diabetic disease	Chitosan improve bio-availability besides intestinal sorption resulted low blood glucose.
Chitosan-DNA nanostructures	Entrapped plasma DNA carrier	Chitosan-nanostructures guard entrapped plasma DNA from nuclease filth.
Chitosan conjugated complex	Encapsulated conjugate delivery	Highly targets tumor cells.
Chitosan doped drugs	Fluorescein drug gets effectively delivery	Potent drug delivery on epithelial cells of ocular mucosa.

Table 1.
Certain chitosan formulations/composites used for controlled drug delivery [1].

owing narrow particle size 1 and 10 nm also achieved by means of reverse micellar medium to be used for effective drug delivery. Methodical encapsulation of other conjugates in chitosan matrix is viable to get prominent NPs using surfactant dissolved in an organic medium to yield reverse micelles with advanced R&D view to be utilized in macromolecule delivery. Fractional conjugation of PEG at alkaline conditions resulted self-aggregated amide linked soluble composites which ensnare insulin drug via electrostatic interactions with residual cell proteins. **Table 1** shows a selection of studies on the utilization of chitosan NP composites for drug delivery systems.

2.3 Safe chitosan composites

Advanced alterations in chitosan frameworks to yield assorted nanostructured matrix can proffer myriad biomedical/pharmaceutical applications. Chitosan is comparatively safe due to abundance in nature from renewable sources, biodegradable and biocompatible nature; yet unmodified chitosan usage is limited by virtue of huge hydrophobicity and high viscosity which tends facile coagulation with proteins at high pH. Despite many limitations still chitosan is vulnerably amazing matrix for drug delivery purpose. These nanocomposites derived from chitosan are prestigiously advantageous over conventional counterparts due to enormous surface area and supplementary features procured through blending/grafting/impregnation particularly for clinical utility. Chitosan matrix plays vital role in tissue engineering and fabricated composite of this bio-polymer acts as good bone implant materials. Research efficacy of chitosan derived hybrid/composite materials cater many challenges and own myriad functionality in S&T [17].

2.4 Sculptured formulations

Biological polysaccharides like chitin/chitosan proved the most excellent model toward the formulation with hydroxyapatites via tempted apatite nucleation which resulted HA crystal growth by virtue of super-active —C = O, —OH and —NH_2 functionalities. Quality and quantitative share of staples chitosan besides other vital factors like presence of inorganic ions, pH and temperature governs effective mineralization of hydroxyapatites. α,β-Chitosan staple-HA crystals of chondroitin sulfate scaffolds are obtained as major bio-matrix hosting embed guests practicable for native physicochemical and hierarchical controls anticipated in bio-mineralized tissue replacement materials. Chitosan has elevated empathy for charged octacalcium phosphate/OCP acting as a herald to enamel, dentine and bones compartmentally formulated crystals own orientated/alternated hydrated-apatite coating akin hydroxyapatite/HA. Nano-hydroxyapatite yields via chitosan-gelatin networking surface are facile to modulate under amicable conditions of adjustable charges, templates density; temperature [1, 17]. Chitosan is facile for intervening poly-anionic linear 1-4-R-D-galacturonosyl/methyl esters/1-2-R-L-rhamnopyranosyls impart apatite-formation due to innate carboxyl functionality which undergoes mineralization along with catalytic heterogeneous apatite nucleation. Skilled chitosan-apatite hydrogels with increased strength obtained via rationally designed egg-box skeleton with Ca^{2+} cations [18].

2.5 Step up chitosan matrix: to get rid native limiting traits for myriad utility

Ever growing scientific decisive exigent it needs stepping-up adoptable chitosan matrix to overcome native restraining traits and to avail its innumerably challenging applications. Prime barriers in chitosan utility imparted due to certain restrictive characters viz.; reduced mechanical stability/strength, weak crystallinity and low solubility in water as well as in organic solvents ultimately constrained practical utility. Hence, raw chitosan seek vital formulations/modifications performed under doping, blending, grafting and impregnation methodologies which enhance limiting features establishing enough shelf-life for its advanced applications. Mechanical stability gets achieved by manipulating environmental factors and processing conditions viz.; temperature, chemical/ionic stabilizing agents in fabrication of chitosan based matrixes [19].

Multifunctional composites/hybrids/matrixes fabrication put global R&D inputs in modern S&T developments. Chitosan-based materials due to inherent biocompatibility, biodegradability and mucoadhesiveness are used in numerous biomedical applications, including prolong/control release of drugs/cell/genes, cartilage/bone-tissue scaffolds, wound dressings, blood anticoagulants, and space filling implants [20]. Crude chitosan as diverse semi-solid structure undergoes facile and assorted alterations/modifications in mild conditions like at lower pH (than pK_a 6.3), yields non-Newtonian, shear-thinning fluid. Such formulations own good mucoadhesiveness due to cationic nature as imparted by free —OH/—NH_2 interaction with mucin by hydrogen/electrostatic forces, thus acts as suitable excipient for buccal, nasal, ocular and vaginal dosage [1]. Chitosan formulations have shown penetrative enhanced and active transport via epithelium layer encloses tight junctions [1]. Chitosan show high susceptibility to environmental factors and processing conditions like heating and freezing thus impose stressful degradation of its skeleton.

Rather, variable molecular weight, polydispersity, controlled deacetylation degree, purity and % moisture determines degradation/splitting of β-1,4-glycosidic/depolymerization and *N*-acetyl/deacetylation link cleavages resulted decrease molecular weight and raised deacetylation degree. Strong intermolecular interactions of chitosan interchain crosslinking modify its skeleton, leads irreversible loss in physicochemical properties [20].

2.6 Factors affecting chitosans stability

2.6.1 Purity

Chitosan is available in many grades of purity depending on molecular weight, and deacetylation degree. Chitosan manufacturing methods greatly responsible for different qualities and properties with resulted corresponding deviations. Further specifications are frequently curtailed and mislead its utility features. While chitosan recoveries from sources engross demineralization, deproteinization and decoloration which imparts certain impurities, like ashes, heavy metals, and protein causing complex dissolution and impede preparations. Chitosans purity affects biological immunogenicity/biodegradability also alters solvent solubility and mechanical strength/stability while, microbiological contamination enhances its enzymatic hydrolysis/degradations. So chitosan based material formulations seeks contaminants free and high quality extra [21].

2.6.2 Polydispersity

Assorted molecular weight distributions in preparation of chitosan matrix is viable for significant physicochemical and biological features like hydrophilicity, viscosity, water-uptake ability, biodegradability, and mucoadhesion. Based on original resources and corresponding preparation methodologies the commercial grade chitosan own average *molecular weight of* 10–100,000 kDa and estimated via, osmometry, light scattering, NMR, viscometer and chromatographic techniques. These measurements needs properly validation, since molecular weight of chitosan differs based on the applied technique. Degree of deacetylation can decrease molecular weight, uniformity, polydispersity index (between 0.85 and 1.15 own good polymer homogeneity) and proper functionality of chitosan products. High molecular weight chitosan is comparatively more thermal stability than low MW stuffs. Moreover, numerous factors viz.; strong acids/alkali, high temperature, mechanical shear, irradiation found to influence the molecular weight of chitosan. High pressure homogenization, wide shearing, centrifugation often decrease M_W and responsible for the fluctuations in polydispersity index. The compression force drawn in tablet formation is accountable for heat generation so manipulates M_W distribution.

2.6.3 Prototype degree of deacetylation

The degree of deacetylation viably accounts ratio of glucosamine to *N*-acetylglucosamine units, whereas glucosamine and *N*-acetylglucosamine distribution all along its elongated chain illustrates its characteristics pattern of deacetylation. Chitosan's deacetylation degree is managed via adaptable time and temperature involved in de-*N*-acetylation according to specification parameter around 70–95%. Deacetylation conditions demonstrated characteristic P_A varies from block to random vital to accurately define its degree of de-*N*-acetylation and P_A which are crucial factors for attaining desired physicochemical/biological functions. Low DD induces sharp inflammatory response due to quick degradation rate, while high DD grounds minimal inflammation and *in-vitro* minor affinity to enzymes. P_A and DD found to manipulate biodegradability via homogenously distributed acetylation yields inferior enzymatic degradation. Studies revealed alteration in DD found to influence hydrolytic and thermal capacity of chitosan derived materials as more de-*N*-acetylated slower acidic hydrolysis during storage. Highly deacetylated chitosan owe less porosity, lesser hydrophilic, and more photo-sensitive for degradation thus restricts degradation in acidic conditions. P_A significantly collies with charge density and affects chitosans solubility owing identical M_W and DD as block of acetylation/deacetylation which gets aggregated in acidic environment and hinders its dissolution process.

2.6.4 % Moisture

Basically chitosan is hygroscopic thus capably forms hydrogen bonding via inherent —OH and —NH_2 functionalities which affects relative humidity i.e., moisture content depending on storage in surrounding temperature independent of DD or M_W. While water-uptake capacity/hydrophilicity of chitosan based materials found to decrease with enhancement of degree of de-acetylation as absorbed water plays critical role in solid formulations by affecting concern flow properties and compressible tensile strength. Fluctuated moisture level in chitosan derived material alters physicochemical and mechanical properties as dehydration decreases crushing strength besides augmented friable disintegration. Higher moisture content imparts faster pronounced damage via hydrolysis and limits chitosans applicability, thus it need to optimize/reformulate or moisture. Swelling index testing is used to investigate water-uptake ability changes upon long-term storage of chitosan materials for both semi-solid and solid formulations.

2.7 Chitosan stability affecting factors

2.7.1 Environmental conditions

The environmental factors are very crucial for sensitive chitosan and it can be stored in closed containers at temperatures of 2–8°C particularly in ascertaining shelf-life. The extra stable chitosan materials provide reliable quality of chitosan. The crucial environmental parameters like humidity and temperature alters physicochemical properties and applications of chitosan [22].

2.7.2 Humidity

Moisture i.e., ambient relative humidity of chitosan strongly controls transport followed Fickian process, as high humidity is viable for an anomalous diffusion kinetic. Humidity > 60% is responsible for water penetration more intensively via chitosan chains, thus % moisture increased significantly resulted plasticizing/swelling and prolong storage results hydrolytic damage besides alter physicochemical and biological characters. Ambient humidity 75% is viable for greater swelling of chitosan and liable for better and faster release of drug as a carrier. Overall, undue hydration at elevated humidity fades mucoadhesiveness of chitosan based drug carriers due to "dilution" of functionality accessible for mucin adhesive interactions. Suitable humidity conditions are important for storage of solid chitosan formulation products as rate of hydration found extensive at high RH. Also proper air-tight containers are advisable in order to protect hygroscopic products against interfering environmental humidity.

2.7.3 Heat/temperature

Heat/temperature variably affects the water content in chitosan-based materials. Elevated temperature > 40°C can origins major moisture loss/dehydration that gives decreased hardness and mechanical strengths. Atmospheric temperature found to influence degradation and hydrolysis rate of chitosan matrix, mainly in liquid and semi-solid phases. However, chain hydrolysis is not observed in the chitosan storage at 5°C, thus verifies storage in a refrigerator at 2–8°C.

2.7.4 Acidic dissolution/processing

Hydrolysis is problematic in pharmaceutics due to its dissolution in diluted acids as scheduled in chitosan-based formulations. Acid catalyzes splitting of polymer chains/linkages depending upon acid type/concentration, treatment time, and temperature besides decreases its average molecular weight, viscosity and weaken mechanical strength. Chitosan hydrolysis is performed usually under specific organic acids like lactic, formic, lactic, and mineral acids hydrochloric. Faster chain damage observed if lower DD chitosan which own extra porosity and electrostatic repulsion between protonated NH_2 which promotes penetration of acid inside its flexible skeleton. Chitosan gets decomposed in aqueous acetic acid at 5°C and intrinsic viscosity under specific solvent and temperature direct affects polymers average molecular weight. Mark-Houwink exponent explores alterations in chitosans specific conformation via amplified chain length indicated framework as $\alpha = 0$ consign compact sphere and $\alpha = 0.7$ refer random coil while $\alpha = 2$ own rigid ceilings helpful in determining average molecular weights.

2.7.5 Sterilization

Sterilization eliminate/remove/kill/deactivate all living forms and species present on, or chitosan matrix employed for drug administered owing high microbiological purity. Chitosan formulations/materials are usually sterilized either physically or chemically leading irreversible alteration in its structural features and function via many techniques like filter sterilization, steam sterilization, dry heat, ethylene oxide activated, and γ-radiation exposures. Sterilization of chitosan gels can be achieved through saturated steam found to cause chain/linkage scission resulting about 50% decrease in viscosity and loss of molecular weight. Similarly, autoclaving sterilizes chitosan films and reduces inherent tensile strength via interchain crosslinking of amino groups which ultimately weaken polymeric solubility. There are no significant changes observed in structure of chitosan through autoclaving also its molecular weight is unaffected once steam sterilized prior to autoclave. Gamma/γ-irradiated sterilization cause significant chain scissions owing low water sorption capacity and decreases molecular weights depending on doses of irradiation and polymer chain rearrangements. Epoxide exposure causes minor changes in morphology and physicochemical features of chitosan restricted to surfaces. Ultraviolet light displayed degradation of chitosan via free radical formations that can destruct amino/hydroxy groups.

2.7.6 Thermal heating

Heat is frequently employed in chitosan-based formulation which is responsible to change polymeric properties, like aqueous solubility, viscosity, and its appearance. Chitosan decompose/damage is on heating at rising temperature and span of heating. The first stage degradation occurs at 30–110°C due to evaporation of the residual water and second thermal damage in chitosan skeleton observed over temperature range 180–340°C. Differences in glass transition temperatures resulted diverse increasing molecular weight chitosan. Third stage degradation is viewed at 470°C due to subsequent weight loss of chitosan. Thus, chitosan matrix is advisable to heat up to temperatures below glass transition temperature with unaltered physicochemical features. Still the gentle heating is necessary to dissolve chitosan in acidic solution since overheating cause's polymer discoloration and depolymerization which eventually change rheological properties. The added/doped drugs, plasticizer or additives in chitosan matrix reduces its glass transition temperature. Thus excipients employed in chitosan formulations preparation are doped at the temperature 120 and 170°C. Thermal decomposition alters electrostatic charges resulting higher hydrolysis and accelerated aggregations.

2.7.7 Lyophilization

Freeze-drying or lyophilization is established drying way in which chitosan is dried by sublimation of ice which is advantageous to prevent aggregations. *Lyophilization* of chitosan can feasibly improve physicochemical stability of colloidal microparticulate in formulated delivery products over extended time periods with better physicochemical stability. Yet, lyophilization impose stress on unmodified chitosan and damage its polymeric chains via weakening inter/intramolecular hydrogen bonding and hydrophobic interactions [23]. This brings negatively effect on viscosity, zeta potential, and water-uptake ability of chitosan formulations.

2.8 Strategic improvements in chitosan products

Chitosan polymer owes poor stability over time renders unsuitable/inapplicable in the pharmaceutical products. Thus, effortful researches are done to improve the stability of chitosan formulations without affecting its chain damage as shown in **Figure 1** [24].

2.8.1 Stabilizing agents

Chitosan is very susceptible to physicochemical degradation upon storage thus it needs to apply proper excipients so as to improve stability of chitosan-based

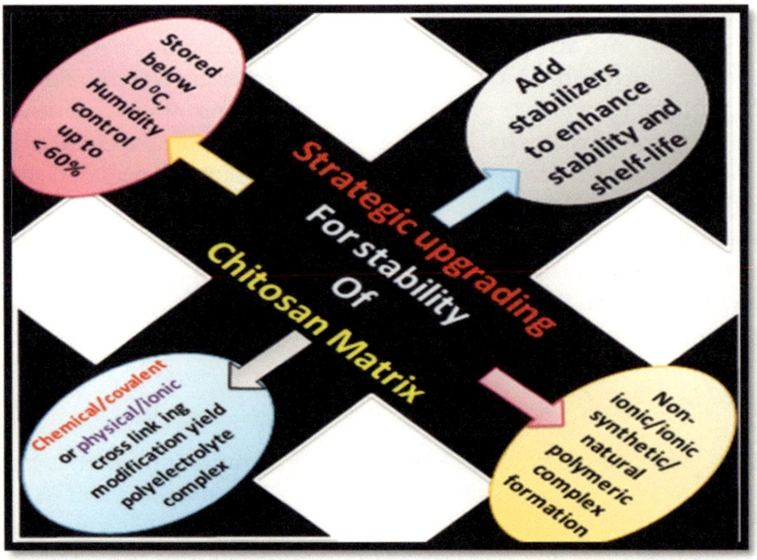

Figure 1.
Strategic upgrading for stability of chitosan matrix to improve their productivity.

systems. Dry heat exposure/steam sterilization have own remarkable consequence on properties and performance chitosan formulations. Thus developed assorted stabilizing additives to protect chitosan during thermal processing and sterilization. Polyols like mannitol, sorbitol, and glycerol as *the stabilizing agents* in chitosan formulations prior to autoclave protects hydration layer around its skeletal surface through interchain hydrogen bonds which markedly slow down its degradation besides protects viscosity and thermogelling properties. Chitosan microparticles obtained via ionotropic gelation followed by tripolyphosphate sodium crosslinking in polyethylene glycol shown stabilized zeta potential and prevent aggregation over long span. Added stabilizer reduces the particles electrostatic charge and led to aggregation after long storage.

Stabilizing polyols like disaccharides act as water replacement agents and interact via hydrogen bonding replaced water besides highly viscous sugar hinder labile materials from disruption via freezing. This plasticizer added to chitosan formulations can manipulate water-uptake capacity/hydrophilicity and mechanical strength that consequence prolonged and controlled drug release profile. Metal ions like Zn(II) are also added to enlarge colloidal stability of chitosan polyelectrolyte formulations due to imparted stabilization attributed to coordinate bondings tune to morphological alteration and swelling properties.

2.8.2 Blends/hybrids

Chitosan blends/hybrids are obtained via nonionic additives responsible to improve physicochemical properties than both constituting agents. Mixtures of chitosan with starch, poly-vinyl alcohol, pol-ethylene oxide and polyvinylpyrrolidone enhances material stability of resultant blended matrix. Chitosan undergoes specific interactive blendings as achieved via hydrogen, ionic and/or dipolar interference with residual component's dependent miscibility display decreased moisture sensitivity. Blended modification controls water-uptake capacity and own higher thermal degradation compared to pure chitosan and proportionated addition. Such blending controls and improves thermal/hydrolytic stability besides biodegradability which conveys resistive enzymatic degradation.

2.8.3 Physicochemical crosslinking

Chitosan skeleton undergoes significant modifications via various physicochemical crosslinking. Added agents-chitosan blends forms chemical crosslinking via covalent bonding and physical crosslinking via ionic bonding. Chemical guard the physicochemical stability of chitosan since gelation is irreversible higher stability is achieved through more covalent bonds besides hydrogen/hydrophobic bondings. Rather chemical crosslinking changes biological properties and limits practical pharmaceuticals utility. Crosslinking level markedly influenced swelling ability, mucoadhesiveness, acidic stability of microparticulates and color alterations.

While ionic/physical crosslinking bridges negative charged components like citrate, sulfate, phosphate groups binds faster onto cationic chitosan which prevents protonation of chitosan amino groups to yield polyelectrolyte complexes. Physical modification is simple, facile and mild requires no catalysts and extra purification in contrast to chemical crosslinking. Chitosan-acyclovir crosslinking are achieved via solvent change with sodium citrate salting shown better physical stability to drugs viable for its controlled release. Microparticulate delivery systems derived from chitosan are strongly depends on surface electrostatic charges that gets altered upon storage, so strategies are developed to prevent aggregation and corresponding zeta potential changes. Improved microparticles stability is attributed due to added non-ionic stabilize polyoxyethylene sorbitan sodium monooleate over ionic crosslinkings.

Despite the great potential of using chitosan in drug delivery or tissue engineering systems, its poor long-term stability is a substantial drawback in the scaling-up of chitosan pharmaceutical applications. Upon storage, chitosan undergoes gradual chain degradation followed by destruction of its functional groups which as a consequence leads to irreversible loss of its physicochemical properties. Both intrinsic (degree of deacetylation, molecular weight, purity, and moisture level) and extrinsic factors (environmental storage conditions, thermal processing, sterilization, and processing involving acidic dissolution) are acknowledged as crucial parameters affecting the stability of the chitosan-based formulations. To improve chitosan stability, several strategies (addition of the stabilizing agent during the preparation process, blending with hydrophilic polymer, and use of ionic or chemical cross-linkers) have also been reported. As there are no universal principles to preserve chitosan-based products upon storage, preformulation studies and selection of the most proper storage conditions are essential to provide their maximal stability.

2.9 Chitosan formulation for water treatments

2.9.1 Case study of defluoridation/fluoride removal

Chitosan is facile to sophisticated biological and physicochemical adaptations in its inherently flexible skeleton so as to yield novel composite/blend which own huge and widespread applications than its contemporary cellulose [1–4, 25]. Qualitative and quantitative framework changes in chitosan matrix can offer highly facile industrial grade suitable formulations/fabricated products and solutions have provided ever demanding exertion in water and wastewater treatments. Biosorption of fluoride from water onto fabricated chitosan-graphite novel composite is illustrated.

2.9.2 Synthetic scheme of chitosan doped bio-composite

Chitosan gelling obtained via dissolution in acetic acid at mild acidic pH gets intertwined via invasive hydrophilic/phobic interaction and induces impulsive entanglement in self-standing microsphere hydrogel as shown in **Figure 2**.

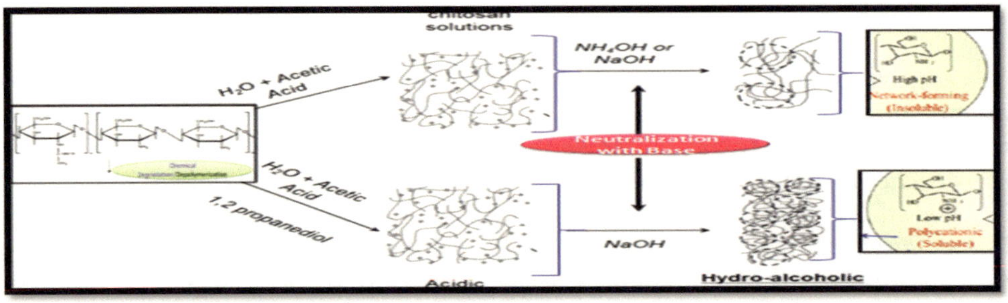

Figure 2.
Preparation scheme for chitosan gelling (scale-5 mm).

This phenomenon followed by coagulation in alkaline solution which subsequently yields viscous droplets/bids are further treated with graphite resulting desired fabricated chitosan doped graphite composites FCDGC. Chitosans hydrogel on drying or evaporation causes dramatic shrinkages in its pore size to impart porosity owing elevated specific surface area. This micro-porosity is attributed to space zones of contacts between chitosan fibrils though impregnation onto —OH/—NH_2 by doped graphite surfaces as shown in **Figure 3**.

2.9.3 Fluoride biosorption onto FCDGC

The fluoride anions are facile to sorbed by FCDGC due to diffusive interaction via weak intermolecular forces as bridge to connect fluoride onto activated surfaces of adsorbent that ultimately enhanced sorption capacity. Although, amine/hydroxyl groups of FCDGC plays vital role in bio-sorption of fluoride, however, other functionalities also affect fluoride sorption may be due to surface complexation, physic-sorption and chelation affinity suitable for scavenging fluoride at pH 6.5. Further, decreased in fluoride sorption in above pH > 6.5 is interpreted due to ligand-exchange between fluoride and counter anion hydroxide coordinated on immobilized FCDGC. This developed FCDGC displayed a surface controlled

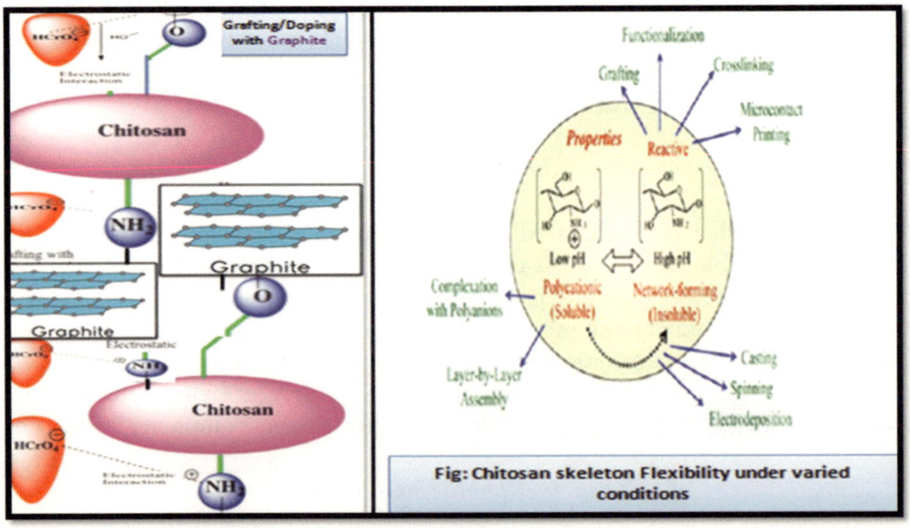

Figure 3.
Mechanistic layers of graphite and chitosan yielding doped bio-composite illustrated inset flexible change in chitosan skeletons.

monolayer sorption of fluoride with interactive heterogeneous distribution and diffusion to cationic sites/surface of adsorbents. The mechanistic view of fabricated chitosan doped graphite biocomposite FCDGC and insight for elevated bio-sorption is depicted in **Figures 4** and **5** respectively.

2.10 Chitosan smart materials: viable for copious challenges

Micro to nanoscale smart materials own certain multi-dimensional switching characters ever utilized in advancement of science due to repetitive self-healed auto-altering of temperature, pressure, heat, electricity and light environments [26]. Smart approaches aid designing of rational chitosan materials that convey salutary purpose and usage including piezoelectric, shape memory polymers, thermo-responsive polymers, photomechanical stuff, self-healing materials and thermoelectric resources, besides hydrogels, nanoassemblies, super-active surfaces and bio-conjugates. Biopolymer chitosan based fabricated materials owe auto-adjustable mechanical strength and depth by virtue of intrinsic cenotaph ability along with inventive status on stress disburses. This chapter sum up revolutionary growth and interest to supply various vital applications. Typically designed parameters can formulate stylish compatibility in chitosan skeleton that can able to counter slight changes in adjustable environment. Chitosan permits rapid changes/transitions via synergic merge of individual characters of foreign components in resultant matrix so as to yield enviable properties which hold uniqueness to carry innovative functionalizations. Thus, smart and responsive nanoassembly/nanoparticles are sensibly designed by means of advanced nanobiotechnology. Chitosan based 3D hydrogels are able to absorb enormous water while sustains invariable stability to bear volume phase/gel-sol phase transitions due to adoptable alterations. These hydrogels possess various physic-chemically stimulated characters viable to bring various responses. Smart surface/interface own two-phase intermolecular force discontinuity, and thus attain very unique high energy point. Chitosan based

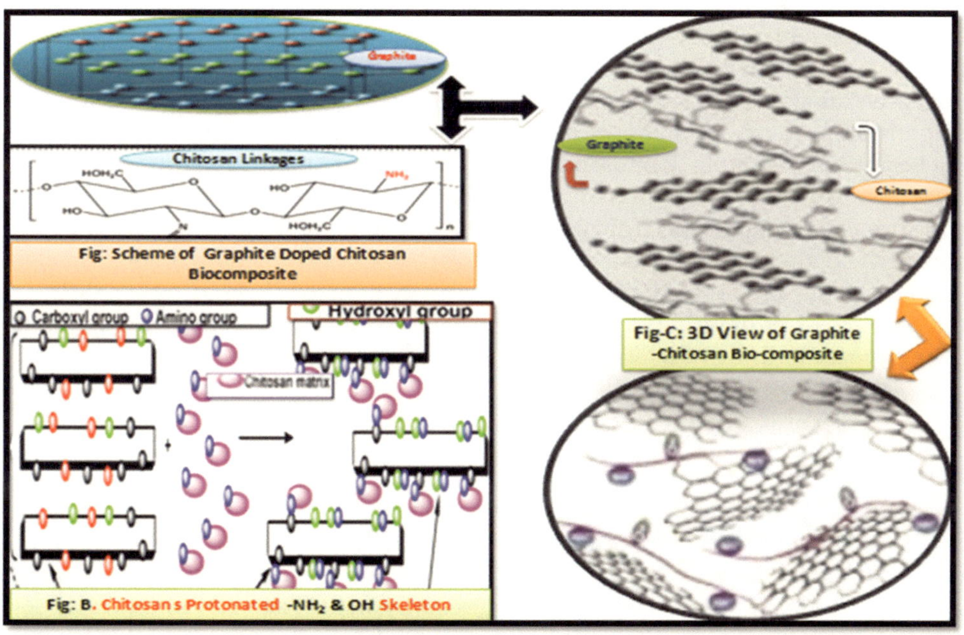

Figure 4.
Mechanistic view of fabricated chitosan doped graphite composite/FCDGC.

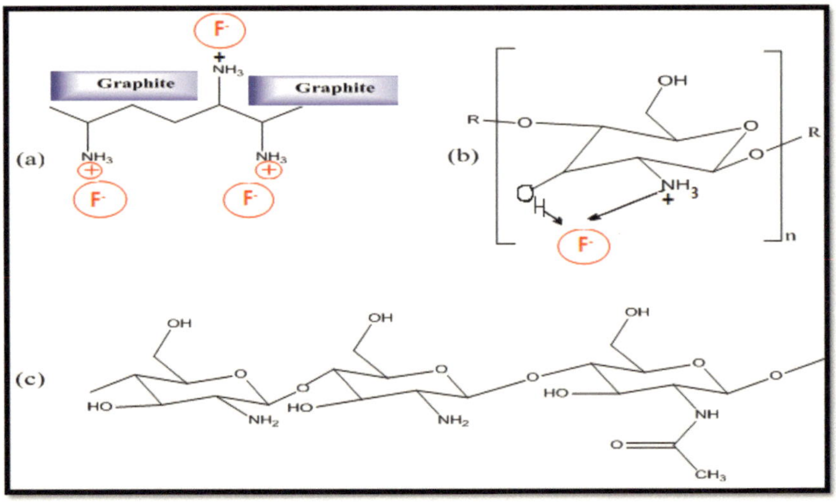

Figure 5.
View of fluoride interactive adsorption mechanism on of FCDGC active sites.

"smart surface/interface are designed with dynamically controllable properties to be utilized for assorted biomedical applications viz.; adsorption of biomolecules, tissue engineering and bio-separations besides biocompatible materials at the biotic/abiotic interface.

Chitosan based stimulus-responsive smart nanofibers are obtained for their 'on–off' reversible switching actions owing exclusive advantageous of nanodimensions imparting peculiar features like huge surface area, high porosity and enhanced external stimuli sensitivity, besides simplistic bulk manipulations in their resultant skeleton. Such stimulus-responsive smart chitosan derived nanofibers own dynamic and reversibly tunable structures with potential 'on–off' actions crucial for efficient delivery of drug/cell/gene in assorted medical applications. Smart bio-conjugates based on chitosan possess superior features and unique properties to original chitosan and added versatile new values due to generation of nanoscale switching. Bio-conjugate smart polymeric matrix are employed for enormous applications viz.; proteins affinity separations, enzyme bioprocesses, drug/cell/gene carrier, diagnostics purpose, biomarkers, biosensors and cell cultured tissue engineering besides DNA motors. Chitosan based shape-memory materials comprise the capability to change from a temporary to memorized permanent shape via peripheral stimulus responses. Such shape-memory polymers (SMPs) can act as a cheap and efficient alternative to well-known metallic shape-memory alloys due to facile manufacturing and easy programming. Thermally induced chitosan based SMPs own self-repairing/rewritable features which owe special weightage in development of environmentally benign technologies [1–4, 27].

3. Summary

Chitin/chitosan matrix is favored to derive sophisticated formulations to be used for its promising risk-free functions in assorted fields like clinical, biomedical, and pharmacological, besides designing/fabrication of advanced nano/biotechnological smart materials. Chitosan template infringes many organic as well inorganic cationic and anionic materials in its flexible/amicable framework so as to get hybrids, hydrogels, composites, coatings, films and nanostructures owing ample utilities in modern S&T. Chitosan still seek methodical R&D to dope/blend interactive fillers/

dopants or additives which can offer improved formulations with split wide open revolutionary and advanced applications.

Nanotechnology integrated chitosan derived smart formulations possess wide, multitasking and thematic portfolio in nanotechnology way from "biosensor/biomarker matrixes" to artificial atoms called "quantum dot". These advanced technologically designed characteristics chitosan materials initiated new modality owing innovative utilities including DNA/RNA/cell/gene nano-carriers, quantum dots for disease diagnosis/therapeutics besides tissue scaffold designing as templates and devices for benefit of man and nature.

Acknowledgements

The author is thankful to the Head, Department of Chemistry, R.T.M. Nagpur University, Nagpur, for laboratory facilities and to the Vice Chancellor, Nagpur University, Nagpur, for the sanction of a research project under University Research Project Scheme, No. Dev/RTMNURP/AH/1672 (9) dated 24 September 2016.

Author details

Rajendra Sukhadeorao Dongre
Department of Chemistry, RTM, Nagpur University, Nagpur, MS, India

*Address all correspondence to: rsdongre@hotmail.com

References

[1] Dongre RS. Biological Activities & Application of Marine Polysaccharides. Vol. 1. Croatia. ISBN: 978-953-51-2859-5: In-Tech Open; 2017. pp. 181-206. DOI: 10.5772/65786

[2] Dongre RS. Chitosan-derived synthetic ion exchangers: Characteristics and applications. In: New Trends in Ion Exchange Studies. ISBN: 978-1-78984-247-0. Vol. 1. Croatia: In-Tech Open; 2018. DOI: 10.5772/intechopen.78964

[3] Dongre RS. Introductory chapter: Multitask portfolio of chitin/chitosan: Biomatrix to quantum dot. In: Chitin-Chitosan Myriad Functionalities in Science and Technology. Vol. 1. Croatia. ISBN: 978-1-78923-406-0: In-Tech Open; 2018. DOI: 10.5772/intechopen.71146

[4] Jain T, Kumar S, Dutta PK. Chitosan in the light of nano-biotechnology: A mini review. Journal of Biomedical Technology and Research. 2015;**1**(1):101-107

[5] Bănică F-G. Chemical Sensors & Biosensors: Fundamentals & Applications. UK: John Wiley Sons; 2012. p. 176. ISBN: 978-0-470-71067-8

[6] Jianc H, Su W, Caracci S, Bunninc TJ, et al. Optical wave guiding and morphology of chitosan thin films. Journal of Applied Polymer Science. 1996;**61**:1163-1171

[7] Burkatovskaya M, Tegos GP, Swietlik E, Demidova TN, P Castano A. Use of chitosan bandage to prevent fatal infections developing from highly contaminated wounds in mice. Biomaterials. 2006;**27**:4157-4164

[8] Kurita K. Chitin & chitosan: Functional biopolymers from marine crustaceans. Marine Biotechnology. 2006;**8**(3):203-226

[9] Mourya VK, Inamdar NN. Chitosan-modifications and applications: Opportunities galore. Reactive and Functional Polymers. 2008;**68**(6):1013-1051

[10] Sashiwa H, Shigemasa Y. Chemical modification of chitin & chitosan 2: Preparation & water soluble property of N-acylated or N-alkylated partially deacetylated chitins. Carbohydrate Polymers. 1999;**39**(2):127-138

[11] Kumar Dutta P, Dutta J, Tripathi VS. Chitin and chitosan: Chemistry, properties & applications. Journal of Scientific & Industrial Research. 2004;**63**:20-31

[12] Peter MG, Domard A, Muzzarelli RAA. Advances in Chitin Science. Vol. IV. Potsdam, Germany: Universität Potsdam; 2000. p. 395. ISBN: 3-9806494-5-8

[13] Hudson SM, Jenkins DW. Chitin & chitosan. In: Encyclopedia of Polymer Science and Technology-1. 3rd ed. New York: Wiley Interscience. DOI: 10.1002/0471440264.pst052

[14] Dutta PK, Ravikumar MNV, Dutta J. Chitin and chitosan for versatile applications. Journal of Macromolecular Science-C, Polymer Review. 2002;**42**(3):307-354

[15] Noipa T, Ngamdee K, Tuntulani T, Ngeontae W. Cysteamine CdS quantum dots decorated with Fe^{3+} as a fluorescence sensor for the detection of PPi. Spectrochimica Acta—Part A: Molecular and Biomolecular Spectroscopy. 2014;**118**:17-23

[16] Yan J-J, Wang H, Zhou Q-H, You Y-Z. Reversible and multisensitive quantum dot gels. Macromolecules. 2011;**44**(11):4306-4312

[17] Venkatesan J, Kim S-K. Chitosan composites for bone tissue engineering, an overview. Marine Drugs. 2010;**8**(8): 2252-2266. DOI: 10.3390/md8082252

[18] Falini G, Fermani S. Chitin mineralization. Tissue Engineering. 2004;**10**(1-2):1-6. DOI: 10.1089/107632704322791646

[19] Hardison D, Deepthike HU, Pathirathne T, Wells MJ. Temperature-sensitive microcapsules with variable optical signatures based on incorporation of quantum dots into highly biocompatible hydrogel. Material Chemistry. 2008;**18**(44):5368-5375

[20] Sá-Lima H, Caridade SG, Mano JF, Reis RL. Stimuli-responsive chitosan-starch injectable hydrogels combined with encapsulated adipose-derived stromal cells for articular cartilage regeneration. Soft Matter. 2010;**6**(20):5184-5195

[21] Thongngam M, McClements DJ. Influence of pH, ionic strength & temperature on self-association & interactions of sodium dodecyl sulfate in the absence & presence of chitosan. Langmuir. 2005;**21**(1):79-86

[22] Chen Y, Yao R, Wang Y, Chen M, Qiu T, Zhang C. CdS QDs-chitosan microcapsules with stimuli-responsive property generated by gas-liquid microfluidic technique. Colloids & Surfaces B: Biointerfaces. 2015;**125**:21-27

[23] Li X, Yang Z. Chitosan-decorated calcium hydroxide microcapsules with pH-trigger release for endodontic application. Journal of Materials Chemistry B. 2015;**3**:8884-8891

[24] Kayan DB, Koçak D, İlhan M. The activity of PAni-chitosan composite film decorated with Pt nanoparticles for electrocatalytic hydrogen generation. International Journal of Hydrogen Energy. 2016;**41**(25):10522-10529

[25] Corazzari I, Nisticò R, Turci F, Faga MG, Franzoso F, Tabasso S, et al. Advanced physico-chemical characterization of chitosan by means of TGA coupled on-line with FTIR and GCMS: Thermal degradation and water adsorption capacity. Polymer Degradation and Stability. 2015;**112**:1-9

[26] Prata AS, Grosso CR. Production of microparticles with gelatin and chitosan. Carbohydrate Polymers. 2015;**116**:292-299

[27] Wu D, Delair T. Stabilization of chitosan/hyaluronate colloidal polyelectrolyte complexes in physiological conditions. Carbohydrate Polymers. 2015;**119**:149-158